METRIC MATH:
THE
MODERNIZED
METRIC SYSTEM (SI)

CONTEMPORARY UNDERGRADUATE MATHEMATICS SERIES
Robert J. Wisner, Editor

THE NATURE OF MODERN MATHEMATICS, Karl J. Smith
MATHEMATICS FOR THE LIBERAL ARTS STUDENT, SECOND EDITION, Fred Richman, Carol Walker, and Robert J. Wisner
INTERMEDIATE ALGEBRA, Edward D. Gaughan
ALGEBRA: A PRECALCULUS COURSE, James E. Hall
COLLEGE ALGEBRA, Edward D. Gaughan
TRIGONOMETRY: CIRCULAR FUNCTIONS AND THEIR APPLICATIONS, James E. Hall
INTRODUCTION TO SYMBOLIC LOGIC, Karl J. Smith
MODERN MATHEMATICS: AN ELEMENTARY APPROACH, THIRD EDITION, Ruric E. Wheeler
A PROGRAMMED STUDY OF NUMBER SYSTEMS, Ruric E. Wheeler and Ed R. Wheeler
FUNDAMENTAL COLLEGE MATHEMATICS: NUMBER SYSTEMS AND INTUITIVE GEOMETRY, Ruric E. Wheeler
MODERN MATHEMATICS FOR BUSINESS STUDENTS, Ruric E. Wheeler and W. D. Peeples
FINITE MATHEMATICS: AN INTRODUCTION TO MATHEMATICAL MODELS, Ruric E. Wheeler and W. D. Peeples
ANALYTIC GEOMETRY, James E. Hall
INTRODUCTORY GEOMETRY: AN INFORMAL APPROACH, SECOND EDITION, James R. Smart
MODERN GEOMETRIES, James R. Smart
AN INTUITIVE APPROACH TO ELEMENTARY GEOMETRY, Beauregard Stubblefield
GEOMETRY FOR TEACHERS, Paul B. Johnson and Carol H. Kipps
LINEAR ALGEBRA, James E. Scroggs
ESSENTIALS OF ABSTRACT ALGEBRA, Charles M. Bundrick and John J. Leeson
AN INTRODUCTION TO ABSTRACT ALGEBRA, A. Richard Mitchell and Roger W. Mitchell
MODULES: A PRIMER OF STRUCTURE THEOREM, Tom Head
INTRODUCTION TO ANALYSIS, Edward D. Gaughan
DIFFERENTIAL EQUATIONS AND RELATED TOPICS FOR SCIENCE AND ENGINEERING, Robert W. Hunt
THE ANALYSIS AND SOLUTION OF PARTIAL DIFFERENTIAL EQUATIONS, Robert L. Street
A PRIMER OF COMPLEX VARIABLES WITH AN INTRODUCTION TO ADVANCED TECHNIQUES, Hugh J. Hamilton
CALCULUS OF SEVERAL VARIABLES, E. K. McLachlan
PROBABILITY, Donald R. Barr and Peter W. Zehna
THEORY AND EXAMPLES OF POINT-SET TOPOLOGY, John Greever
AN INTRODUCTION TO ALGEBRAIC TOPOLOGY, John W. Keesee
EXPLORATIONS IN NUMBER THEORY, Jeanne Agnew
NUMBER THEORY: AN INTRODUCTION TO ALGEBRA, Fred Richman

METRIC MATH: THE MODERNIZED METRIC SYSTEM (SI)

JAMES R. SMART
San Jose State University

BROOKS/COLE PUBLISHING COMPANY
MONTEREY, CALIFORNIA

A Division of Wadsworth Publishing Company, Inc.

ISBN: 0-8185-0126-X
L.C. Catalog Card No.: 73-93954
Printed in the United States of America
2 3 4 5 6 7 8 9 10

PREFACE

Measurement is one of the principal connecting links between mathematics and everyday life. The metric system is a scientific system of measurement that can be used in elementary mathematical applications. The advantages of the metric system over the customary English system of measurement used in this country are numerous enough to be impressive. Compared to the English system, the metric system is simpler and quicker to use, more logical, and more scientific. Also, it is used by more countries than is the English system. These advantages should become apparent throughout the text as more information about the metric system is provided.

This book is for those who want to understand the metric system and who want to be able to use it with confidence. Its purpose is to present material and to suggest additional experiences that will help the reader in acquiring the following skills and understandings, which are important in the use of the modernized metric system:

1. An adequate understanding of the definitions and concepts of modern metric units of measurement, for the purpose of feeling secure in their use.
2. The ability to change from one unit of measurement to another within the modernized metric system.
3. The ability to work typical measurement problems within the modernized metric system.
4. The ability to think in metric units, rather than translate to another system, and to make reasonable estimates using metric units.

5. The ability to convert, when necessary, to or from the modernized metric system. This skill includes familiarity with rough conversions and also the ability to use more precise and exact conversion factors.
6. An acquaintance with the historical development of the modernized metric system.

Within ten years, use of the metric system should be far more common than use of the English system. As the modernized metric system becomes the primary system of measurement in the United States, items 2 through 4 become increasingly important and the fifth item less and less important. Working within the metric system is of paramount importance. For people who have used the English system for many years, however, the use of rough conversions will aid them in gaining familiarity with metric units and in acquiring skill in estimating with metric units as readily as with English units.

The material in this book is designed for those who wish to use the metric system in conjunction with mathematics and mathematical measurements. No particular technical knowledge in science is assumed. Definitions of metric units are given as simply as possible, with scientific terms explained whenever practical. No specific mathematical prerequisites beyond arithmetic are necessary.

Most of the material included here has been used for several years in university-level content courses in mathematics. Students in these classes include both elementary and secondary teachers of mathematics. Many of the topics covered have also been included by the author in speeches at professional meetings.

The book is designed for either classroom or individual use. In either case, the mathematics laboratory-type activities are essential, because the necessary skills cannot be developed through reading alone. Actual work with measurement instruments and with real objects is necessary. With the use of mathematics laboratory activities, the text contains enough material for at least 15 periods of instruction at the college level.

Specific uses of this book include, but are not limited to, a supplement to content courses in college mathematics, especially those for elementary or secondary teachers; a separate one-unit module at the college level; a text for in-service, extension, or summer courses for mathematics teachers; a reference for individual study by interested adults; a supplement for high school mathematics courses; and a supplement to courses with mathematics content, such as some homemaking, automobile mechanics, and science courses.

The comprehensive nature of the book makes it easier to adapt to the needs of the various groups of readers mentioned. For example, the entire subject of conversion between measurement systems can be omitted without loss of continuity. For some readers, computational skill with the metric units may be an important objective, whereas, for other readers, understanding of activities involving actual use of the metric units in practical measurements may be much more important.

Chapter 1 is a historical introduction. Chapters 2 and 3 explain the modern metric units of length and the derived units of area and volume. Chapter 4 concerns metric units of mass. Additional units of the modernized metric system are considered in Chapter 5. Students with a greater need to apply metric measurements to science will find the last chapter of great value.

Although a large number of metric units are mentioned, the average person uses only a few of them with much frequency. For example, units of length and area are needed much more often than units of luminous intensity or amount of substance. Moreover, among the units of length, a few units, such as the meter, centimeter, and kilometer, are used more frequently than the others. Fortunately, becoming familiar with the most common metric units is both easy and interesting.

For their helpful criticisms and suggestions in the preparation of the manuscript, I thank George W. Bright of Northern Illinois University, Arthur Hallerberg of Valparaiso University, Vincent Hawkins of Toll Gate High School in Warwick, Rhode Island, Gerardus Vervoort of Lakehead University, Nancy Whitman of the University of Hawaii, and series editor Robert J. Wisner of New Mexico State University. I would also like to thank the efficient staff of Brooks/Cole Publishing Company for their creativity and cooperation.

James R. Smart

CONTENTS

THE DEVELOPMENT OF THE MODERNIZED METRIC SYSTEM

1

In popular usage, the expression **metric system** is used to refer to the modern international system of measurement based on such units as the meter, kilogram, and liter. Actually, the technical name is **Systèmе International d'Unites,** universally abbreviated **SI** and known in the United States as the **International System of Units.** SI, adopted in 1960, is somewhat different from the metric system as originally conceived and as the units were originally defined. It is the International System, rather than the original metric system itself, that everyone now needs to understand. To avoid confusion because of the established popular usage, however, the expressions "International System," "modernized metric system," and just "metric system" are used interchangeably after this short historical introduction.

Before the Metric System

The system of measurement historically used for most everyday measurements in the United States is called the **English system** or the **customary system** of measurement. Its common units, such as the foot, the pound, and the quart, together with their parts such as the inch, ounce, and pint, have now been defined technically in terms of the metric system. In some sense, then, the United States has legally been on the metric system for over a century. The recent development is the plan for the modernized metric system to be the predominant system for everyday measurements.

One of the major advantages of the metric system as compared with the customary system is that the multiples and submultiples of the basic metric units involve powers of ten, which leads to simplicity of changes of units because of our decimal numeration system. As compared with the ancient concept of common fractions, the notation of decimals is relatively recent, being credited to Simon Stevin, a Dutch mathematician, in 1585. The invention of decimals was followed by the development of computing devices and techniques to take advantage of the simplified system of numeration. These devices included the desk calculator, logarithms, and slide rule. It should not be surprising that decimals were soon proposed for use in an improved system of measurement.

Gabriel Mouton, vicar of St. Paul's Church in Lyons, France, proposed in 1670 a decimal system of measurement. For the basic unit of the system, he suggested the length of an arc of one minute of a great circle of the earth. In other words, the unit was $\frac{1}{60}$ of $\frac{1}{90}$th of the distance from the North Pole to the Equator. A different early proposal suggested the length of a pendulum beating once per second at 45 degrees latitude as the basic unit. Both proposals involved standards from nature or the earth itself rather than standards such as the foot or the yard, which were originally measurements of the human body. The use of standards from nature, instead of body measurements, is a significant advantage of the metric system, and is one of the reasons the metric system is considered more scientific than the English system.

The Original Metric System

The metric system in its original form was developed by a commission of the French Academy of Sciences at the request of the National Assembly of France. The major ideas appeared in a report made to the French National Assembly in 1791. The original system contained several standard units that need to be explained briefly here, although the original definitions are no longer applicable.

The **meter** (also spelled metre) was, and still is, the basic unit of length. The intention was to construct a metal bar whose length would be one ten-millionth of the distance from the North Pole to the Equator. The actual arc measured (over a seven-year period ending in 1798) was from Dunkerque, France, to a place near Barcelona, Spain, as shown symbolically in Figure 1.1. Then a certain fraction of this distance was defined as the meter. When improvements in methods of measuring showed that the arc had been measured inaccurately, the meter was redefined as the distance between two marks on a metal bar

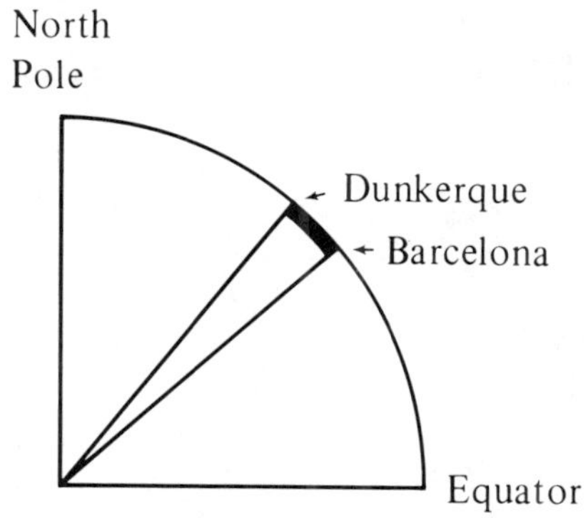

Figure 1.1. Original arc for metric system.

constructed from the original measurements. The resulting metal bar was established as the **prototype**, an original model to be used as a standard to which other models must conform and then to be duplicated for use in various countries. Multiples and submultiples are related to the meter by powers of ten. Thus, the centimeter is .01 meter and the kilometer is 1,000 meters. For all types of metric units, names of multiples were assigned Greek prefixes and names of submultiples were assigned Latin prefixes. Additional prefixes have been added periodically.

A second metric unit established was the **gram**, a unit of mass. There is a technical difference between mass and weight, to be discussed in Chapter 4, but in practical usage English units of weight are equated to units of mass. Unlike the English system, the original metric system established a basic connection between units of mass and units of length, since the gram was defined as the mass of one cubic centimeter of water at its temperature of maximum density. Figure 1.2 represents in two dimensions the actual size of a container for one gram of water. Other units of mass are defined in terms of the gram, using positive or negative powers of ten. For example, a kilogram

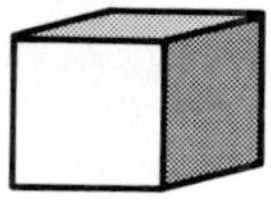

Figure 1.2. Gram of water.

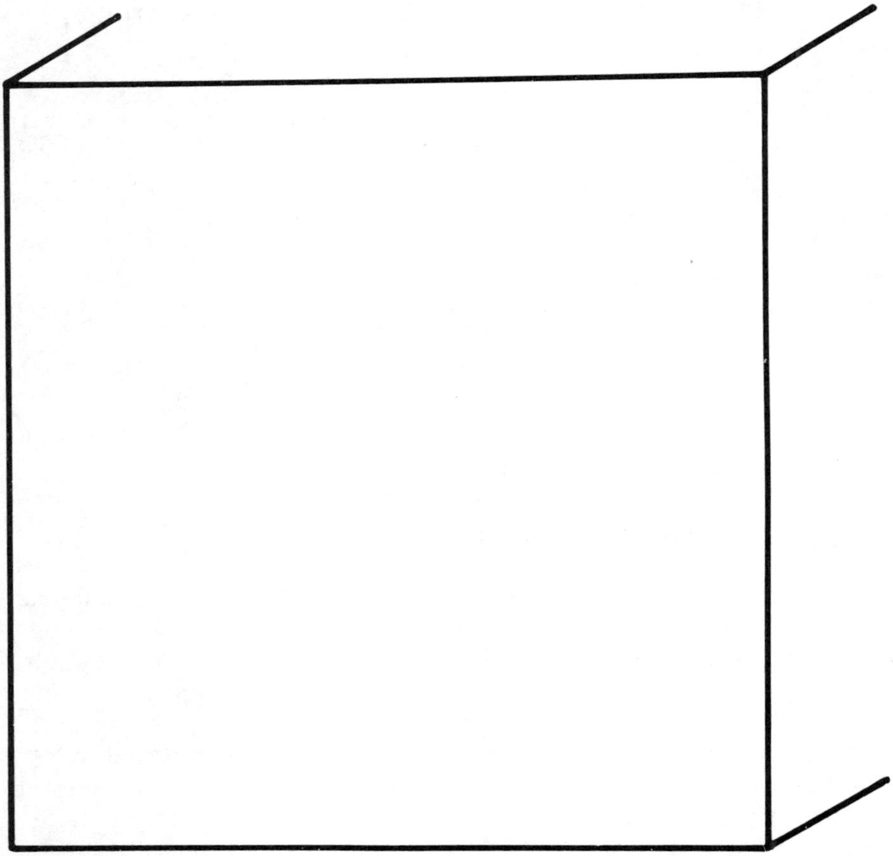

Figure 1.3. Face of one-liter container.

is 1,000 grams.

The third original metric unit was the **liter**, a measure of liquid capacity, defined as the capacity of 1,000 cubic centimeters. Figure 1.3 is a two-dimensional diagram of the actual size of a container with a capacity of one liter. Other defined units included the **are** (pronounced as air), a land measure equal to a square 10 meters on each side, and a **stere**, the name given to one cubic meter, particularly when used to measure piles of material such as cordwood. The units described above are the units of the original metric system, not of the present modernized metric system that grew out of it.

A Universal System

The metric system became the legal system in France in 1795, replacing a system that contained such units of length as the pole, pied, and league. The older system, based on measurements handed down from the time of the Roman occupation, had grown to include many variations. This older system was reinstated as a second legal system in 1812, but, finally, in 1840 the metric system became the only legal system in France. Since that time its use has spread widely. For example, the metric system was adopted in Spain in 1849, Portugal in 1852, Mexico in 1862, Germany in 1870, Switzerland in 1875, Sweden in 1879, Russia in 1900, and China in 1914.

Although it is rather commonly believed that the United States has only recently been concerned with adopting the metric system, this is not true. The history of the metric system in the United States, however, is unlike its history in most other countries.

The various American colonies had adopted English units, but standards differed from one locality to another. The need for a uniform national system of weights and measures was recognized by the framers of the Constitution. In fact, the United States Congress, in 1790, requested Thomas Jefferson, who was then Secretary of State, to make a proposal. Jefferson's first suggestion was to make uniform and permanent the existing English System. But he also proposed an interesting alternative plan that was actually a decimal system of measurement which retained customary names for English units.

As a few English standards become increasingly common in the various states, the Congress again asked a Secretary of State, John Quincy Adams, to make a report. In 1821, after four years of study, Adams presented a careful report. While recognizing the theoretical merits of the metric system, he nevertheless seriously questioned the wisdom of trying to adopt a completely new system of measurement.

At the national level, little had yet been accomplished. In 1830, the Senate directed the Secretary of the Treasury to compare standards of measurement at the various custom houses. As a result of the problems revealed by this study, uniform standards based on the English system were adopted by various state legislatures, although the Congress never officially endorsed the U.S. customary system. Prior to the Civil War, an acceptable degree of uniformity of standards had been achieved without the centralized control so apparent in most other countries.

Although there was resistance to changing in daily usage from the familiar English units to an unfamiliar and more foreign system, actually the United States did not lag behind in cooperating with other countries in the development of the metric system. The use of metric units was made lawful (although not mandatory in everyday affairs) by an act of Congress in 1866. Most English units were then defined legally in terms of metric units. The United States was one of seventeen countries participating in a Metric Convention in 1875, the purpose of which was to set up carefully defined metric standards and to make arrangements for the future development of the metric system. The 1875 agreement provided for an International Bureau of Weights and Measures, with headquarters at Sèvres, France, a suburb of Paris. This arrangement provided a place both for research and for storage of original prototypes.

In 1890 the United States received prototype meters Nos. 21 and 27 and prototype kilograms Nos. 4 and 20. These prototypes were careful copies of the original standards and were to be used to make additional copies for use throughout the nation. Then in 1893 the yard and pound were redefined in terms of metric units. The metric system was legalized for trade in Great Britain in 1897, although it had been permitted earlier. By about 1900, 41 nations had adopted the metric system and 35 nations were using it as their predominant system of measurement. In 1959 the United States and Great Britain agreed to a new common definition of the yard and pound in terms of metric units, to replace the slightly different definitions previously adopted independently by the two countries. The British plan to convert to the metric system during the period 1965–1975 left the United States as one of the few nations still using the English system in its everyday affairs.

In recent years, significant changes in the definitions of basic metric units and the definitions of additional basic units have resulted in a much more usable system for scientific measurements. The system adopted in 1960, and since further modernized, is called (in the United States) the **International System of Units,** abbreviated SI, for **Système International d'Unites.** This system was adopted by the General Conference of Weights and Measures, the organization made up of delegates from all member nations adhering to the original Metric Convention. SI should be considered as an improvement and an extension of the metric system.

Exercise Set 1

1. Name two major differences mentioned between the English and the metric systems of measurement that show advantages of the metric system.

2. What was the basic unit suggested by Gabriel Mouton for his decimal system of measurement?
3. In what country was the original metric system developed?
4. What was the original definition of the meter?
5. Define the original basic unit of mass in the metric system.
6. Define the original unit of liquid capacity in the metric system.
7. The stere was primarily used to measure what objects?
8. When did the use of metric units become lawful in the United States?
9. What is the official name of the modernized metric system?
10. The recent conversion of what nation to the metric system was influential in the plans of the United States for conversion?
11. Which was longer, Mouton's proposed standard of length or the original meter?

INTERNATIONAL SYSTEM UNITS OF LENGTH

2

2.1 DEFINITIONS OF SI UNITS OF LENGTH

The Basic Unit of Length

The fundamental unit of length in the International System, and the unit that gave the metric system its name, is the **meter**. This word comes from the Greek word *metron,* meaning a measure. As explained in Chapter 1, the original meter was defined as one ten-millionth of the distance from the North Pole to the Equator, along a certain line of longitude passing through France. Today, in the modernized metric system (SI), the meter is defined in terms of wavelengths of light. A meter is 1,650,763.73 wavelengths in vacuum of the orange-red line of the spectrum of krypton-86. The original metal bar established as a result of the measurement of arc had a disadvantage in that, if the original standard were ever lost, it could never be reproduced. The current definition, using wavelengths, does not have the built-in flaw associated with use of a metal bar as the primary standard, since the meter can now be reproduced in almost any scientific laboratory.

More about the Definition of Meter

A person who wishes can use meters without understanding the definition. A somewhat simplified explanation of how the meter is determined will provide sufficient understanding of the concept to proceed with its use in mathematics.

Krypton, a chemical element discovered in 1898, is an inert gas with the atomic number 36. The nucleus of a krypton atom contains 36 protons. Krypton is used as an element in the mixture in most flourescent lamps and in luminous tubes for signs. Krypton-86 is what is called an **isotope.** Isotopes are atoms of the same element that differ in atomic weight because of a differing number of neutrons in the nucleus. The number 86 indicates the total number of protons and neutrons in the nucleus.

When light passes through a transparent prism, it spreads out in a band of colors called a **spectrum.** No two materials have the same spectrum. Spectrum analysis (using a spectroscope) reveals information about the light source because an excited atom, such as the krypton-86, emits light at each of a number of exact wavelengths. Light produces a series of spectral lines, with gases such as krypton producing clearer lines than solids or liquids.

In the definition of the meter, a multiple of the length of a particular light wave is used. Figure 2.1 illustrates what is meant by one wavelength. The instrument used to count the number of light waves in a meter is called an **interferometer.** A. A. Michelson, an

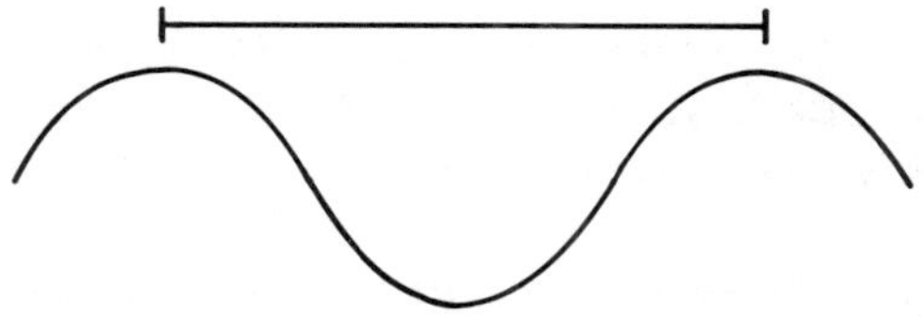

Figure 2.1. Meaning of length of 1 light wave.

American physicist, invented a particular kind of interferometer in 1881 that could be used to count light waves. He managed to count the number of red cadmium light waves in a meter. The definition of the meter was changed in 1960 to depend on the orange line in a discharge of krypton gas consisting of the single isotope 86. This line is sharper than the red cadmium line and hence better for a very precise measurement.

In an interferometer, a half-silvered mirror divides the beam into two equal parts. The light beam traveling over different paths and reconverging cause fixed fluctuations of intensity that physicists call *interference fringes.* The interferometer is an optical-type system producing a visual pattern of these fringes that can be counted. If the different light paths have different lengths, then each wavelength difference means a shift of one interference fringe. In this way, small differences, within a small fraction of a light wave, can be measured.

Multiples and Submultiples of a Meter

Multiples and submultiples of length are all expressed as some multiple of the length of a meter. The simplicity in use of the metric system depends mainly on the fact that all SI units have multiples and submultiples based on powers of ten.

The most frequently used multiples and submultiples of the meter, and possibly the only ones needed by the average person, are the six listed below.

1 kilometer	=	1,000 meters
1 hectometer	=	100 meters
1 dekameter	=	10 meters
1 decimeter	=	0.1 meter
1 centimeter	=	0.01 meter
1 millimeter	=	0.001 meter

At least five of the six names listed have familiar prefixes. Kilo is commonly used in kilowatt to mean 1,000 watts. Deka (deca) is used in decade, deci is used in decimal, cent is used as a coin worth 0.01 dollar, and mil is a unit used in rate of taxation.

The complete system assigns names for various powers from 10^{12} down to 10^{-18}. Multiples and submultiples of the meter are shown in the following list.

1 terameter	=	10^{12} meters	1 Tm	=	10^{12} m
1 gigameter	=	10^{9} meters	1 Gm	=	10^{9} m
1 megameter	=	10^{6} meters	1 Mm	=	10^{6} m
1 kilometer	=	10^{3} meters	1 km	=	10^{3} m
1 hectometer	=	10^{2} meters	1 hm	=	10^{2} m
1 dekameter	=	10 meters	1 dam	=	10 m
1 decimeter	=	10^{-1} meters	1 dm	=	10^{-1} m
1 centimeter	=	10^{-2} meters	1 cm	=	10^{-2} m
1 millimeter	=	10^{-3} meters	1 mm	=	10^{-3} m
1 micrometer	=	10^{-6} meters	1 μm	=	10^{-6} m
1 nanometer	=	10^{-9} meters	1 nm	=	10^{-9} m
1 picometer	=	10^{-12} meters	1 pm	=	10^{-12} m

1 femtometer	= 10^{-15} meters	1 fm	= 10^{-15} m
1 attometer	= 10^{-18} meters	1 am	= 10^{-18} m

The prefixes in the preceding list, such as mega and kilo, are standardized for all SI units, not just for length. However, the spelling of some of the words is not uniform in every country. For example, "deca" is used in SI, but "deka" has been adopted by the National Bureau of Standards. The symbol for micro is the Greek letter mu, written μ, since m and M had already been used for other prefixes. It is important to realize that the symbols given in the list are those in effect in 1973, but that SI, like many other parts of mathematics, is subject to change as conferences are held and new agreements are reached. Note, for example, that punctuation is not used with the symbols, and this is a relatively recent convention.

The pronunciation of multiples and submultiples of SI units is not completely standardized. In general, units with four syllables are pronounced with the first and third syllables accented. But kilometer is pronounced either kil′-o-me′-ter or kil-om′-eter.

The prefixes hecto, deca, deci, and centi are commonly used in measuring length, but generally they are not used with other basic SI units; such SI multiples and submultiples proceed in units of one thousand.

Exponent Notation

The use of SI multiples and submultiples of the meter is often made simpler by employing the notation of exponents. Recall the definitions of integral exponents:

$b^n = \underbrace{b \cdot b \cdot b \cdots b}_{n \text{ factors}}$ for b a nonzero real number and n a positive integer. This notation means that b is used n times as a factor.

$b^0 = 1$ for b a nonzero real number.

$b^{-n} = 1/b^n$ for b a nonzero real number and n a positive integer.

The definition of negative exponents can be used to rewrite submultiples using decimals. Thus, 1 cm = 0.01 m and 1 mm = 0.001 m, so 100 cm = 1 m and 1,000 mm = 1 m.

The simplest instrument in common usage to measure meters

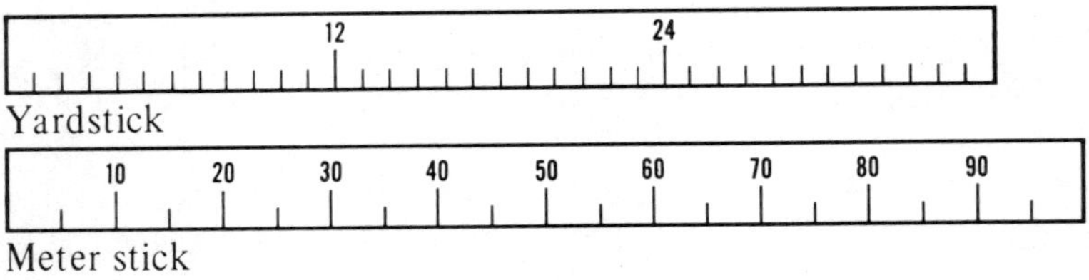

Figure 2.2. Meter stick to scale.

is a **meter stick**, illustrated to scale in Figure 2.2. Rulers are frequently marked in centimeters and millimeters. A 30-cm measuring stick is very common. The reader should have available both a meter stick and a 30-cm stick. An instrument called a **micrometer** (pronounced mi-crom′-eter), pictured in Figure 2.3, may also measure in millimeters or micrometers.

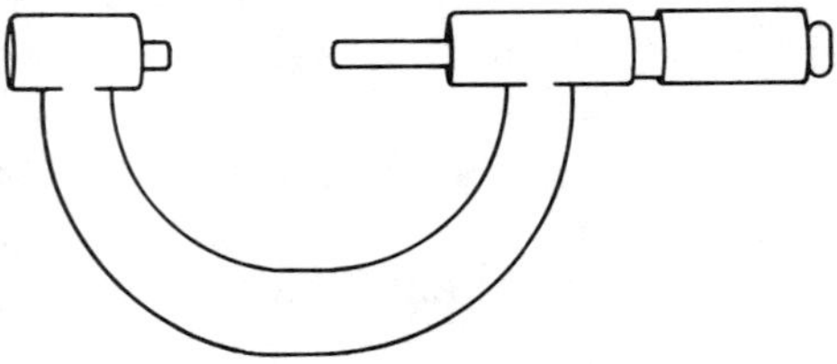

Figure 2.3. Micrometer.

Exercise Set 2

For Exercises 1 through 4, write the number without using exponents.

1. 2^7 2. 3^4 3. 10^5 4. 10^0

For Exercises 5 through 8, write the number without using negative exponents.

5. 3^{-4} 6. 5^{-3} 7. 10^{-3} 8. 10^{-5}

For Exercises 8 through 18, use decimal notation to show how many meters are equal to each given multiple or submultiple.

9. 1 hectometer 10. 1 kilometer
11. 1 megameter 12. 1 gigameter
13. 1 decimeter 14. 1 centimeter
15. 1 millimeter 16. 1 micrometer
17. 1 nanometer 18. 1 picometer

Mathematics Laboratory Activities 1

For activities 1 to 6, use a meter stick.

1. Measure 10 objects as carefully as possible.
2. About how many of your hand spans equal one meter?
3. About how many of your foot lengths equal one meter?
4. About how many pencils laid end to end have a combined length of one meter?
5. How many dollar bills laid end to end have a combined length of one meter?
6. Is the distance from the floor to your waist more or less than one meter?

For activities 7 to 10, use a 30-cm ruler.

7. Measure 10 objects as carefully as possible. Report the results in centimeters.
8. How many pennies are needed to make a stack 1 cm high?
9. Is a thumbtack longer or shorter than 1 cm?
10. How many paper matches, laid side to side, are needed to have a width of 1 cm?
11. Using a ruler marked with millimeters, measure 10 objects as carefully as possible. Report the results in millimeters.
12. Measure the height of three people and record the results in SI units.
13. Check on the length in meters of various Olympic swimming and track-and-field events.
14. Practice using a micrometer marked in SI units, to gain more familiarity in measuring to the nearest millimeter.
15. Examine a set of metric tools in which sockets and wrenches are measured in millimeters. For some particular automobile for which metric tools are needed, determine the sizes most commonly needed.

2.2 MAKING ESTIMATES USING SI MEASUREMENTS OF LENGTH

For anyone who has first learned the English System of measurement, one of the important abilities to acquire is that of estimating distances in the metric system. This is an ability that can be improved with practice, and it contributes to an increased understanding of the SI units.

One aid to estimating is to set up distances to be used as standards. For example, you may know that the distance from your house to the grocery store is 2 kilometers. By comparing other distances with this known one, you may improve your estimates rather quickly. These individual standards may differ from one person to another and are often used in the more familiar system of measurement with little awareness. Useful individual standards include actual body measurements, such as height and length of an arm.

For the meter, a mental image of the meter stick is helpful for making comparisons. For example, the height of a very tall man may be estimated as 2 m because it is possible to visualize two meter sticks placed end to end as being approximately the height of the man. Another useful approximation to a meter is the distance from a door knob to the bottom of the door. For centimeters, visualizing a standard 30-cm ruler and knowing that the diameter of a nickel is approximately 2 cm are two ideas that may be helpful in estimating. For the millimeter, two useful ideas are that a dime is about 1 mm thick and the wire used for a paper clip has a diameter of about 1 mm.

Estimating distances is aided by having specific examples of known lengths that one can refer to. These examples are not just the visualization of a meter stick or a 30-cm scale; it also helps to know that one's span of hand might be a little more than 20 cm, or the length of an average thumb is about 7 cm. These body measurements may then physically be brought into play when estimating the size of a book or of a table; indeed, use of such crutches in estimating is a natural thing to do.

The laboratory-type activities at the end of this section are designed to help develop the ability to estimate with SI units of length. *In no case should there be any translation from one system of measurement to another for these activities.* In each case, the estimates or guesses should be made strictly within the metric system. Then the results should be compared with actual measurements or with information obtained from reference books.

Mathematics Laboratory Activities 2

For activities 1 to 8, estimate your body measurements in SI units; then check. The results of these exercises should be used to help estimate answers to later exercises.

1. length of your arm
2. length of your thumb
3. width of your hand
4. length of your foot

5. size of your waist
6. width of your little finger
7. your height
8. the span of your hand

For activities 9 to 18, make each estimate directly in meters; then check by using a meter stick to measure the particular object chosen.

9. length of a bed
10. height of a friend
11. height of a desk
12. length of a house
13. length of an automobile
14. length of a room
15. height of a room
16. width of a door
17. width of a sidewalk
18. width of a driveway

For activities 19 to 28, make each estimate directly in centimeters; then check by using a ruler marked in centimeters.

19. length of a pencil
20. length of a comb
21. height of a tin can
22. size of a book
23. size of a check
24. diameter of a half dollar
25. width of a pencil
26. height of a cereal box
27. length of a dollar bill
28. width of a dollar bill

For activities 29 to 32, draw segments as near these lengths as possible by estimating; then check by measuring.

29. 3 cm
30. 5 cm
31. 7 cm
32. 13 cm

For activities 33 to 40, make each estimate directly in millimeters; then check by using a ruler marked in millimeters.

33. width of a pencil
34. thickness of 100 sheets of notebook paper
35. thickness of a dime
36. length of a thumbtack
37. width of head of a pin
38. regular postage-stamp size
39. thickness of a book
40. width of typewriter ribbon

For activities 41 to 58, estimate the distances in kilometers between the two locations. Then use an atlas to check the estimate by using a map with a scale in metric units.

41. Where you were born and where you now live
42. St. Louis and where you now live
43. New York and Washington
44. Seattle and San Diego
45. Los Angeles and Chicago
46. Cleveland and St. Louis
47. Phoenix and Atlanta
48. Boston and Miami
49. Detroit and Atlanta
50. Dallas and Minneapolis
51. New Orleans and Houston
52. Des Moines and Denver
53. Rome and Paris
54. Cairo and Ankara
55. London and Bonn
56. Tokyo and Singapore
57. Montreal and Vancouver
58. Honolulu and Sydney

2.3
CHANGES AMONG SI UNITS OF LENGTH

Each line of the list of multiples and submultiples in Section 2.1 can be written in two ways. For example, since 10^3 m = 1 km, 1 m = 10^{-3} km. Each side of the equality was multiplied by 10^{-3}. In other words, kilometers may be expressed in terms of meters, or meters may be expressed in terms of kilometers.

Two examples illustrate changing from one metric unit of length to another. In many cases, computation of this type can be done mentally, since the digits in the answer are unchanged, but the decimal point is in a different position.

Example 1

Express 352 km in meters.

$$\begin{aligned} \text{Since } 1 \text{ km} &= 10^3 \text{ m}, \\ 352 \text{ km} &= 352 \times 10^3 \text{ m, or} \\ & 352{,}000 \text{ m.} \end{aligned}$$

Example 2

Express 157 nm in meters.

$$\begin{aligned} \text{Since } 1 \text{ nm} &= 10^{-9} \text{ m} \\ 157 \text{ nm} &= 157 \times 10^{-9} \text{ m, or} \\ & 0.000000157 \text{ m.} \end{aligned}$$

For cases like Examples 1 and 2, a few texts use a symbol such as $\doteq$ rather than = for "equals in measure."

Scientific Notation

These examples illustrate the need for a standard way of writing answers when numbers are very large or very small. Scientific notation is commonly used (more often by other scientists than by mathematicians) as a convenient notation because it provides a means of comparing numbers at a glance. A number is in scientific notation when it is expressed as the product of a number between (and including) one and (but not including) 10, times a power of 10. This notation lends itself well to work in the SI because of the computation with powers of ten. Thus in Example 1 the answer in scientific notation is 3.52×10^5, and the answer in Example 2 is 1.57×10^{-7}.

When scientific notation is not used, SI notation often uses spaces. Thus, the answers to Examples 1 and 2 could be written 352 000 m and 0.000 000 157 m.

Example 3

Express 3.47 m in centimeters.
$3.47 \text{ m} = 3.47 \times 10^2 \text{ cm}$.
The substitution is based on the statement
$1 \text{ m} = 10^2 \text{ cm}$.

Example 4

Express 3.47 m in kilometers.
$3.47 \text{ m} = 3.47 \times 10^{-3} \text{ km}$.

The power of the decimal notation, and an advantage of the metric system, is illustrated by the simplicity of change within the system and the variety of ways in which an answer may be written. For example, 347 cm, 34.7 dm, 3.47 m, 0.347 dam, 0.0347 hm, and 0.00347 km are all names for the same measurement. In each case, the names contain the same sequence of digits.

Suggestions to Eliminate Common Errors

Examples 1 through 4 suggest generalizations that will eliminate some common errors in computational techniques with SI units. These generalizations are not meant as rules to learn without understanding. The number will be larger after changing from a larger to a smaller unit of measurement. In Example 3, 347 is greater than 3.47, since the measurement is changed from a larger unit (meters) to a smaller unit (centimeters). The number of units will be smaller after changing from a smaller to a larger unit of measurement. In Example 4, 0.00347 is smaller than 3.47, since the measurement is changed from a smaller unit (meters) to a larger unit (kilometers).

When the answers to problems involving a change of units are written in ordinary decimal notation, the number of places the decimal point in the answer differs from the position of the decimal point in the original number equals the exponent of 10 in the statement of equivalence. If the exponent is positive, as in Example 1, the new position is to the right of the original position. If the exponent is negative, as in Example 2, the new position is to the left of the original position. Again, these are not meaningless rules to be memorized.

Previous examples have illustrated changing to and from meters. Changing between any two SI units of length is almost as simple. The examples can be shortened by the use of mental computation.

Example 5

Change 275 cm to decimeters.

$$\begin{aligned} 275 \text{ cm} &= 275 \times 10^{-2} \text{ m} \\ &= 275 \times 10^{-2} \times 10 \text{ dm} \\ &= 275 \times 10^{-1} \text{ dm} \\ &= 27.5 \text{ dm.} \end{aligned}$$

Example 6

Change 37 μm to centimeters.

$$\begin{aligned} 37\ \mu\text{m} &= 37 \times 10^{-6} \text{ m} \\ &= 37 \times 10^{-6} \times 10^{2} \text{ cm} \\ &= 37 \times 10^{-4} \text{ cm} \\ &= 0.0037 \text{ cm or } 3.7 \times 10^{-3} \text{ cm.} \end{aligned}$$

Examples 5 and 6 can be worked mentally, with some practice, although each step has been written here for complete understanding.

While working the following set of exercises, remember that the kilometer, meter, centimeter, and millimeter are frequently used in practical applications, whereas use of the very large and very small related units is more specialized. The exercises contribute to familiarity with a variety of units of length and to increased skill in changing among multiples and submultiples of the meter. Mental computation is encouraged. The person who needs only a minimum acquaintance with the system may limit his work to the most common units: kilometer, meter, centimeter, and millimeter.

Exercise Set 3

For Exercises 1 to 6, write each statement so that one meter is expressed in terms of the other unit.

1. 10^2 m = 1 hm
2. 10^6 m = 1 Mm
3. 10^{-1} m = 1 dm
4. 10^{-2} m = 1 cm
5. 10^{-3} m = 1 mm
6. 10^{-6} m = 1 μm

For Exercises 7 to 12, express each measurement in meters.

7. 43 dam
8. 75 km
9. 243 mm
10. 634 cm
11. 4.38 km
12. 0.392 Mm

For Exercises 13 to 18, change 32 m to the SI unit specified.

13. km
14. Mm
15. dm
16. cm
17. mm
18. μm

For Exercises 19 to 24, change 0.452 m to the SI unit specified.

19. km
20. dm
21. mm
22. cm
23. dam
24. hm

For Exercises 25 to 28, change 3.457 cm to the SI unit specified.

25. km
26. mm
27. dam
28. hm

For Exercises 29 and 30, change 0.238 km to the SI unit specified.

29. cm
30. dam

For Exercises 31 and 32, the **Angstrom unit** is an older designation for an extremely small unit used to measure wavelengths. One Angstrom unit is defined as one hundred-millionth of a centimeter.

31. Express an Angstrom unit in terms of a meter.
32. Express an Angstrom unit in terms of a micrometer.

For Exercises 33 and 34, the **light year** is defined by astronomers as the distance light travels in a year. This distance may be expressed in centimeters as 9.46×10^{17} cm.

33. Express the light year in terms of a meter.
34. Express the light year in terms of a kilometer.

2.4 COMPUTATION WITH SI MEASURES OF LENGTH

Practical problems involving SI measurements of length require appropriate operations with decimals. Examples are given, and in these examples measurements are treated as exact. These examples are not meant to be comprehensive but are merely illustrative of some operations encountered in practical situations.

Example 1

On line *AC*, with *B* between *A* and *C*, segment *AB* has a length of 4.3 cm and segment *BC* has a length of 8.9 cm. How long is segment *AC*?
Segment *AC* has a length of (4.3 + 8.9) cm or 13.2 cm.

Example 2

How many pieces of string 3.1 m long could be cut from a ball of string 5 dam long?
1 dam = 10 m, and
5 dam = 50 m.
Because 50 divided by 3.1 is 16 with a remainder less than 3.1, the answer is 16 pieces, with some string left over.

Example 3

For a fixed cost per kilometer, the cost of a 10.7-km road is what percent of the cost of a road 10.3 km in length?
$10.7 \div 10.3 \approx 1.04$; therefore the answer is approximately 104% ($\approx$ means "is approximately equal to").

Example 4

If an automobile averages 91.4 km per hour for 3.6 hours, how far has it traveled, to the nearest kilometer?
(91.4) (3.6) = 329.04, so the answer is 329 km.

Example 5

If a map has a scale of 1 cm : 1 km (1 : 100,000), find the actual length of a street that appears 3.2 cm long on the map.
1 cm corresponds to 1 km, so 3.2 cm corresponds to 3.2 km and the street is 3.2 km long.

In these examples, and elsewhere in applications of measurement, the user of elementary mathematics performs the computation with numbers (the measures). Scientists, on the other hand, customarily carry units of measurement through the operations.

Errors in Measurements

For Exercise Set 4, and other exercise sets in this text involving computation with measurements, assume that the measurements are exact. The purpose of this assumption is to simplify the problems by avoiding considerations of error in measurement.

When measurements are reported as the result of a physical act of measuring, errors are always involved. The mathematician, in working with measurements, must be aware of the seriousness of the error in interpreting the answers he gets in computation with measurements, whether it involves SI units or not. In actual practice, computation with measurements is often performed as if the measurements are exact; then the final answer is rounded as necessary. Additional information on errors in measurement can be found in Smart, *Introductory Geometry: An Informal Approach,* 2nd ed., Brooks/Cole Publishing Co., 1972.

Exercise Set 4

1. What is the combined thickness of two boards if one is 2.5 cm and the other is 1.8 cm thick?
2. On the first day of their trip, the group hiked 16.3 km and on the second day they hiked 12.7 km. How much farther did they hike the first day?
3. Find the perimeter of a triangle if the sides are 26.4, 32.7 and 15.3 m.
4. If a train averaged 83.5 km per hour for 4.7 hours, how far did it travel?
5. How many books, averaging 4.2 cm thick, can be placed on a shelf that is 2 m long?
6. A bed is 2 m long, but a cot is 8% shorter. How long is the cot?
7. In a speed-skating event, the winner completed the 500-meter race in 40.2 seconds. What was her speed per minute, to the nearest meter?
8. In a 100-meter butterfly swimming race, the winning time was 1 minute, 4.7 seconds. What was the speed per minute, to the nearest meter.

For Exercises 9 to 12, assume that a map has a scale of 1 cm : 1 km.

9. What is the actual length of a river that appears 185 cm long on the map?
10. What is the actual length of an airfield runway that appears 1.3 cm long on the map?
11. An object 374 m in length should appear how long on the map?

12. Two cities that are 105 km apart should appear how far apart on the map?

For Exercises 13 to 16, assume that a map has a scale of 2 cm : 1 km (1 : 50,000).

13. What is the actual length corresponding to a length of 3 cm on the map?
14. What is the actual width of a lake that appears 5.3 cm wide on the map?
15. A tunnel that is actually 7.234 km in length should appear how long on the map?
16. A mountain range 451 km in length appears how long on the map?

Mathematics Laboratory Activities 3

1. Check the various types of practical problems involving metric measurements of length found in secondary school mathematics texts.
2. Make up five practical problems, using metric measurements of length, and then solve them.
3. Using an atlas, study various maps with scales given in metric units, measuring distances on the map and finding the corresponding actual distances.
4. Construct a map with a scale of 1 cm : 1 km of a familiar region. If possible, make the actual measurements rather than use another map.
5. Find the velocity of light in SI units and express it in terms of meters per second.

2.5 CONVERTING TO AND FROM SI UNITS OF LENGTH

For years the ability to convert units of measurement from one system to another will be a needed skill, although it should be used only when necessary. A person should not work problems within one system by first converting to another system, working the problem, and then converting back again. But conversion will be necessary in such situations as when the original data are given in the English system, whereas the answer must be reported in the metric system.

Simple Approximate Conversions

Two different but related skills are necessary in converting from one system to another. The first is the ability to make quick but only approximate mental conversions, or at least conversions with a minimum amount of time and computation. The second is the ability to use conversion tables giving more precise or exact conversion factors. In available written materials, this second skill has been emphasized much more than the first. Yet it is the first skill that is perhaps more important for the typical person in dealing with applications of measurement in everyday life.

Acquiring the ability to convert measurements quickly depends on having available some rather rough comparisons that lend themselves to rapid work. Use of these simple comparisons should cut down laborious multiplying and dividing. A visual comparison to scale of a meter, yard, and foot is given in Figure 2.4. Four approximate numerical comparisons involving a meter are summarized below:

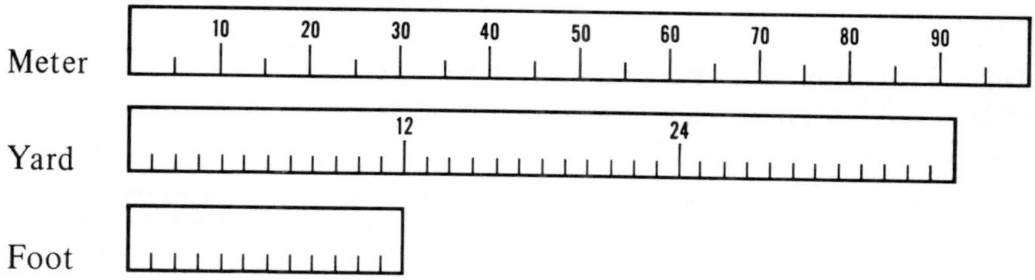

Figure 2.4. Comparison of meter, yard, and foot.

A yard is approximately 0.9 meter.
A meter is approximately 3.3 feet.
A meter is approximately 1.1 yards.
A meter is approximately 39 inches.

Conversions such as 1 m $\approx$ 3 ft $3\frac{1}{2}$ in. were omitted, although their use in mental calculations might be possible. Sometimes easier approximations, such as 40 inches for 1 m, are used instead of those given. Also, a very rough but extremely useful approximation for estimating answers or checking results is that a meter is slightly longer than a yard.

Here are two examples that illustrate approximate conversions by using conversion factors given. Remember that the symbol $\approx$ is used for "approximately equal to."

Example 1

100 yards is about how many meters?

$$\begin{aligned} 1 \text{ yd} &\approx 0.9 \text{ m, and} \\ 100 \text{ yd} &\approx 100\,(0.9) \text{ m} \\ &\approx 90 \text{ m.} \end{aligned}$$

Example 2

50 meters is about how many feet?

$$\begin{aligned} 1 \text{ m} &\approx 3.3 \text{ ft. and} \\ 50 \text{ m} &\approx 50\,(3.3) \text{ ft} \\ &\approx 165 \text{ ft.} \end{aligned}$$

Normally, kilometers and miles are corresponding units for which conversion is sometimes necessary. A scale drawing in Figure 2.5 shows the relative lengths. Two approximate conversion factors are:

A mile is approximately 1.6 kilometers.
A kilometer is approximately 0.6 mile.

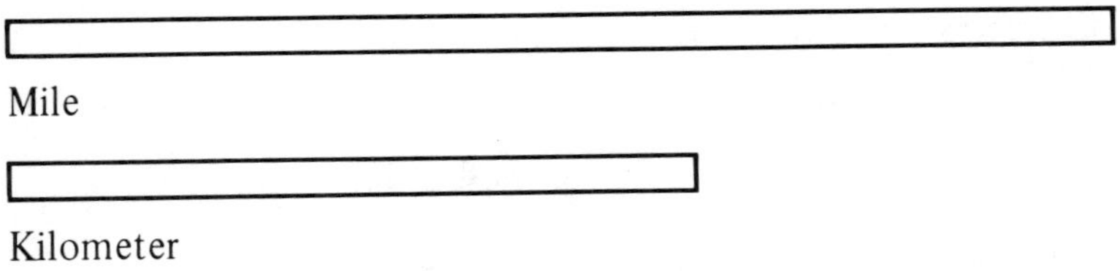

Figure 2.5. Comparison of kilometer and mile.

Two examples use the conversion factors given.

Example 3

A speed limit of 100 km/hr is approximately equivalent to a speed limit of how many miles per hour?

$$\begin{aligned} 1 \text{ km} &\approx 0.6 \text{ mile, and} \\ 100 \text{ km} &\approx 100\,(0.6) \text{ miles} \\ &\approx 60 \text{ miles.} \end{aligned}$$

Then 100 km/hr $\approx$ 60 mph.

Example 4

Two cities that are 92 miles apart are approximately how many kilometers apart, to the nearest kilometer?

$$\begin{aligned} 1 \text{ mile} &\approx 1.6 \text{ km, and} \\ 92 \text{ miles} &\approx 1.6 \times 92 \text{ km} \\ &\approx 147 \text{ km.} \end{aligned}$$

Figure 2.6 shows a scale drawing relating the length of the

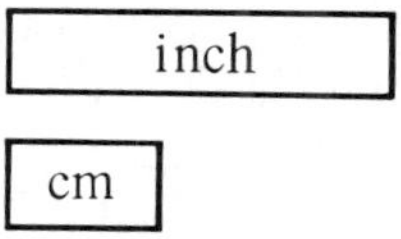

Figure 2.6. Actual length of centimeter and inch.

centimeter and inch. The following four statements give approximate conversions for centimeters.

1 inch is approximately 2.5 cm.
1 foot is approximately 30 cm.
1 centimeter is approximately 0.4 in.
1 yard is approximately 90 cm.

Example 5

A span of 8 inches is approximately how many centimeters, to the nearest centimeter?

$$\begin{aligned} 1 \text{ in.} &\approx 2.5 \text{ cm, and} \\ 8 \text{ in.} &\approx 2.5 \times 8 \text{ cm} \\ &\approx 20 \text{ cm.} \end{aligned}$$

Example 6

Thirty-five centimeters is approximately how many inches, to the nearest inch?

$$\begin{aligned} 1 \text{ cm} &\approx 0.4 \text{ in., and} \\ 35 \text{ cm} &\approx 0.4 \times 35 \text{ in.} \\ &\approx 14 \text{ in.} \end{aligned}$$

The approximate conversions of length illustrated here are adequate for many people. Five very useful approximate conversion factors given previously for changing from English to metric units are summarized in the following list.

If You Know	*Multiply By*	*To Find*
inches	2.5	centimeters
feet	30	centimeters
yards	90	centimeters
yards	0.9	meters
miles	1.6	kilometers

A very convenient source of conversion factors is a plastic card, called the Metric Conversion Card, printed by the United States Department of Commerce. It is for sale at 10¢ from the U.S. Government Printing Office, Washington, D.C. 20402 (Catalog number C13.10: 365).

More Precise or Exact Conversions

When more precise or exact conversions must be made, conversion tables are available. One adequate source is *Units of Measurement, Conversion Factors and Special Tables,* October 1972, United States Department of Commerce. Other sources include books listed in the bibliography, various existing textbooks, and commercially prepared visual aids.

Conversion factors in tables are often identified as exact or approximate. Where they are approximate, results are given correct to a certain number of significant digits. A significant digit is either a nonzero digit or a zero used in a way other than locating a decimal point. For example, the zeros in 23,034.03 are significant. The zero in 20 m is not significant if the meaning is ± 5 m but is significant if the meaning is ± 0.5 m. When scientific notation is used to record the results of a physical measurement, all digits recorded are significant.

Since 1959 one inch has been defined as exactly 2.54 cm, and this exact statement forms the basis for various exact conversion factors. A few of these factors appear in the following table. Obviously, no entry in the table, with the possible exception of 1 in. = 2.54 cm, should be memorized.

	Convert from			
Convert to	Inch	Foot	Yard	Mile
centimeters	2.54	30.48	91.44	160,934.4
meters	0.0254	0.3048	0.9144	1,609.344
kilometers		0.0003048	0.0009144	1.609344

Decimal conversion factors from metric measurement back to English equivalents are all approximate. *Units of Measurement, Conversion Factors and Special Tables,* previously cited, gives these factors correct to seven significant digits. For example, in order to convert from centimeters to inches, multiply by 0.393 700 8. This number is the beginning of the decimal expression for 1/2.54, the reciprocal of 2.54, as should be expected.

Two examples of the use of exact conversion factors follow.

Example 7

Convert 3.2 inches to centimeters.
The multiplication is 3.2×2.54, so the exact answer is 8.128 cm.

Example 8

Express 7 feet in meters.
From the table, the correct conversion factor is 0.3048, so the exact answer is 2.1336 m.

Conversion from one system to another might be considered a necessary evil. The following set of exercises is minimal to avoid an excess of laborious paperwork. Mental estimation, checked by measurement, should be used whenever possible for the approximate conversion. Use a desk calculator for exercises with longer multiplications.

Exercise Set 5

For Exercises 1 to 8, use an approximate conversion factor from among those given in this section to express each measurement to the nearest new unit.

Original Measurement	*Convert Approximately to*
1. 12 yards	meters
2. 17 meters	feet
3. 2.5 meters	yards
4. 31.2 yards	meters
5. 257 miles	kilometers
6. 345 kilometers	miles
7. 15 inches	centimeters
8. 6.4 inches	centimeters

For Exercises 9 to 14, write each of these speeds approximately in kilometers per hour, to the nearest unit.

9. 65 mph
10. 45 mph
11. 25 mph
12. 30 mph
13. 10 mph
14. 15 mph

For Exercises 15 to 18, rewrite each clothing measurement to the nearest centimeter, using approximate conversion factors.

15. 24-inch belt
16. 32-inch inseam
17. size-36 blouse (inches)
18. $15\frac{1}{2}$-inch shirt collar

Use the table of exact conversions to express the measurements in Exercises 19 to 24 in the metric system.

19. 7 inches to centimeters
20. 6 feet to meters
21. 4 yards to meters
22. 25 miles to kilometers
23. 6.2 inches to centimeters
24. 2.61 yards to meters

Mathematics Laboratory Activities 4

For Activities 1 and 2, use a table of conversion factors and a desk calculator.

1. Practice making additional precise changes from English to metric units.
2. Practice making additional precise changes from metric to English units.
3. Find a newspaper account of a beauty contest and write the winner's measurements in centimeters.
4. Write the length and width of a football field in metric units.
5. On a baseball diamond, how far is it from first to second base, and from the pitcher's mound to home plate in SI units?
6. Find the price per SI unit for three articles sold by length (such as rope).

7. In a reference almanac, such as the *Official Associated Press Almanac,* find a list of the highest dams in meters and in feet. Check some of the equivalent measurements, using exact and approximate conversion factors.
8. In a reference almanac, find a list of the longest bridges in meters and in feet. Check some of the equivalent measurements, using exact and approximate conversion factors.

INTERNATIONAL SYSTEM UNITS OF AREA AND VOLUME

3

3.1 INTERNATIONAL SYSTEM UNITS OF AREA

Square Meter, a Derived Unit

The SI unit for area is the **square meter,** with symbol m^2. A square meter is the measure of area of a surface one meter square. Figure 3.1 shows a model of a square meter. The definition of square meter is derived from the basic unit of length in the metric system. For this reason, square meter is called a **derived unit** rather than a basic unit.

Multiples and Submultiples

The three most commonly used relationships with multiples and submultiples of the square meter, and the most important for the average person to be able to use, are:

$$1\ m^2 = 100\ dm^2 = 10{,}000\ cm^2 \quad \text{and}$$
$$1\ dm^2 = 100\ cm^2.$$

A more comprehensive list is provided for those who wish to apply SI units to larger and smaller areas. It is not necessary to memorize the statements. For each equation, the exponent of ten can be determined by reasoning from the list of equivalent lengths given previously in Section 2.1. For example, 1 cm $= 10^{-2}$ m, so

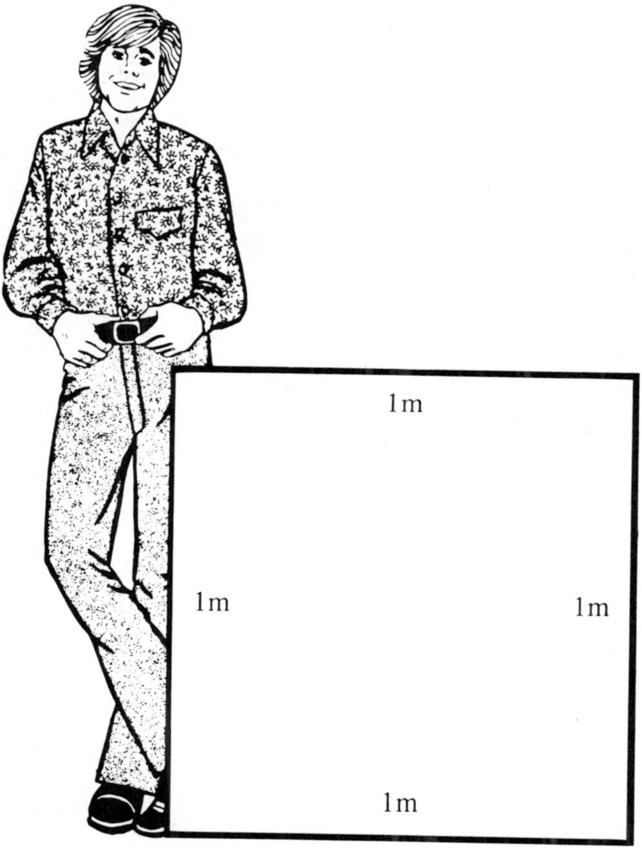

Figure 3.1. Model of 1 square meter.

$1\ \text{cm}^2 = 10^{-2} \times 10^{-2}\ \text{m}^2 = 10^{-4}\ \text{m}^2$. Figure 3.2 shows the actual size of a square centimeter. Powers of ten in the list can be expressed in decimal form if desired. For example, $1\ \text{mm}^2 = (0.001\ \text{m})^2 = 0.000\,001\ \text{m}^2$.

$1\ \text{Mm}^2 = 10^{12}\ \text{m}^2$	$1\ \text{dm}^2 = 10^{-2}\ \text{m}^2$
$1\ \text{km}^2 = 10^{6}\ \text{m}^2$	$1\ \text{cm}^2 = 10^{-4}\ \text{m}^2$
$1\ \text{hm}^2 = 10^{4}\ \text{m}^2$	$1\ \text{mm}^2 = 10^{-6}\ \text{m}^2$
$1\ \text{dam}^2 = 10^{2}\ \text{m}^2$	$1\ \mu\text{m}^2 = 10^{-12}\ \text{m}^2$
	$1\ \text{nm}^2 = 10^{-18}\ \text{m}^2$

The measurement $1\ \text{hm}^2$ is commonly called a hectare (with

Figure 3.2. Actual size of 1 square centimeter.

the symbol ha) when it is used for land measurement. The are, a land measure equal to a square 10 meters on each side, may also be used when dealing with land measurements.

Changes with Area Measurements

Two examples of changing within the metric system of area measurements follow. In many cases, computation of this type can be done mentally, since the digits in the answer are unchanged, but the decimal point is in a different position.

Example 1

Express 156 km^2 in square meters.

$$156 \text{ km}^2 = 156 \times 10^6 \text{ m}^2$$
$$= 1.56 \times 10^8 \text{ m}^2.$$

Example 2

Express 35 cm^2 in square meters.

$$35 \text{ cm}^2 = 35 \times 10^{-4} \text{ m}^2$$
$$= 3.5 \times 10^{-3} \text{ m}^2.$$

Computation with SI Units of Area

Four problems are given to provide examples in which computation involves SI units of measurement of area. The measures are considered as exact.

Example 3

Two small metal parts are to be coated with an expensive alloy. What is the combined surface area to be covered if one surface has an area of 4.7 cm^2 and the second has a surface area of 3.9 cm^2?

$4.7 + 3.9 = 8.6$, so the surface area is 8.6 cm^2.

Example 4

A farm of 200 ha was purchased by a land developer to be divided into lots of 2.4 ha each. How many lots of this size could be formed?
200 divided by 2.4 is 83 with a remainder less than 2.4, so the answer is 83, with some land remaining.

Example 5

Forest fires burned approximately 12.3% of an area of 954 km^2. Approximately how many square kilometers were burned?
954 (0.123) = 117.342, so the answer is approximately 117 km^2.

Example 6

Find the area of a triangular region whose base is 3.5 cm and whose height is 7.4 cm.
The area is $\frac{1}{2}$ (3.5) (7.4) cm^2, which is equal to 12.95 cm^2.

Exercise Set 6

For Exercises 1 to 6, change each of these measurements to square meters.

1. 26 km^2
2. 13 hm^2
3. 42 dam^2
4. 328 cm^2
5. 8.63 dm^2
6. 3.15 mm^2

For Exercises 7 to 12, change the measurement in meters to the unit indicated.

7. 21 m^2 to dam^2
8. 34 m^2 to dam^2
9. 374 m^2 to hm^2
10. 625 m^2 to hm^2
11. 0.025 m^2 to mm^2
12. 0.138 m^2 to cm^2

For Exercises 13 to 18, change the measurement to the unit indicated.

13. 35 km^2 to hm^2
14. 84 cm^2 to dm^2
15. 0.35 cm^2 to μm^2
16. 9.63 nm^2 to cm^2
17. 0.002 km^2 to cm^2
18. 38.91 cm^2 to dam^2
19. Find the combined area of two fields if the first is 7.2 ha and the second is 6.9 ha.
20. Find the area, correct to two significant digits, of a rectangular region if the length is 12.4 m and the width is 7.3 m.
21. Find the area, correct to two significant digits, of a

rectangular region if the length is 0.81 cm and the width is 0.69 cm.

22. Find the area, correct to two significant digits, of a triangular region if the base is 7.5 cm and the altitude is 4.8 cm.
23. Find the height of a triangular region, to the nearest centimeter, if the base is 7.4 cm and the area is 84.26 cm^2.
24. Find the area of a trapezoidal region if the bases are 6.31 cm and 9.47 cm and the height is 1.304 cm.
25. The city annexed a new subdivision 3 km^2, thereby increasing the area of the city by 0.37%. How large is the city now, to the nearest square kilometer?
26. The flood covered 17% of the area of the 2,180 km^2 river valley. How much land was flooded?

For Exercises 27 and 28, use 3.14 as an approximation to π.

27. Find the area, correct to two significant digits, of a circular region if the radius is 1.7 km.
28. A radio station could broadcast to a circular area of 300,000 km^2. What is the radius of the area, correct to the nearest kilometer?

Mathematics Laboratory Activities 5

1. Find the top surface area in SI units of five different flat objects.
2. Find the floor area of your bedroom in metric units.
3. Make a list of the size, in square kilometers, of five different European countries and of five different states.
4. Make up four additional practical problems, using SI measurements of area, and then solve them.
5. Prepare a chart showing a square meter to actual size, with the region partitioned into square decimeters, with one of the square decimeters partitioned into square centimeters and with one of the square centimeters partitioned into square millimeters.
6. In a park or field, mark off a piece of land that is approximately one hectare.

3.2 MAKING ESTIMATES USING SI UNITS OF AREA

In making estimates involving metric units of area, and also in converting from the English system to the metric system of units of area, it is helpful to establish standards for comparison. These standards should be familiar objects that a person can easily visualize in order to make a mental comparison of size when an estimate is needed. For instance, for the square centimeter, one face of a cube of sugar or a small wooden block or die can be used as a model or standard. For the square decimeter, a piece of paper cut to the exact size makes a good model. For the square meter, the surface area of a particular table or a certain number of tiles on the floor may serve as a guide. Some newspapers, such as *The Times* of London, now have the area of a double sheet equal to one square meter. Think of other personal examples to use as an aid in making estimates before continuing.

Although methods of estimating will vary from one person to another, several ideas are offered in the event that no method can be imagined. To estimate the number of square meters in the floor area of a room, three possibilities are:

1. Think about how many meter sticks could be laid across the front and the side and then multiply the results.
2. Think about how many small tables with surface area one square meter would be needed to fill the room completely.
3. Estimate that four of your paces are about 3 meters in length and then pace off the length and width of the room.

To estimate the number of square centimeters of paper in a dollar bill, two possibilities are:

1. Use your thumb width as approximately 2 cm to measure the length and width.
2. Think about how many small pieces of paper 1 cm^2 in size can be placed on the surface.

Mathematics Laboratory Activities 6

For Activities 1 to 8, write estimates directly in square meters; then check your estimates.

1. area of floor of a room
2. area of wall of a room
3. area of a paved driveway
4. area of page of newspaper
5. area of a small rug
6. material needed for dress
7. area of a football field
8. area of tennis court

For Activities 9 to 14, write estimates in square centimeters; then check your estimates.

9. size of front of a book
10. area of a post card
11. size of a check
12. size of a dollar bill
13. top surface area of half dollar
14. top surface area of watch dial

For Activities 15 to 18, write estimates in square millimeters.

15. area of a postage stamp
16. top surface area of a nickel
17. area of head of a pin
18. area of billfold-size photograph

3.3 CONVERTING TO AND FROM SI UNITS OF AREA

Quick Approximate Conversion

The ability, when necessary, to make rapid conversions from one system of area measurement to the other is facilitated by use of the following approximate conversions, which require a minimum of computation. The list should not be memorized.

1 square centimeter equals approximately 0.16 square inch.
1 square inch equals approximately 6.5 square centimeters.

1 square meter equals approximately 1.2 square yards.
1 square yard equals approximately 0.8 square meter.

1 square kilometer equals approximately 0.4 square mile.
1 square mile equals approximately 2.6 square kilometers.

1 hectare equals approximately 2.5 acres.
1 acre equals approximately 0.4 hectare.

Figure 3.3 illustrates the relative size of various square units. Following it are two examples illustrating the approximate conversions.

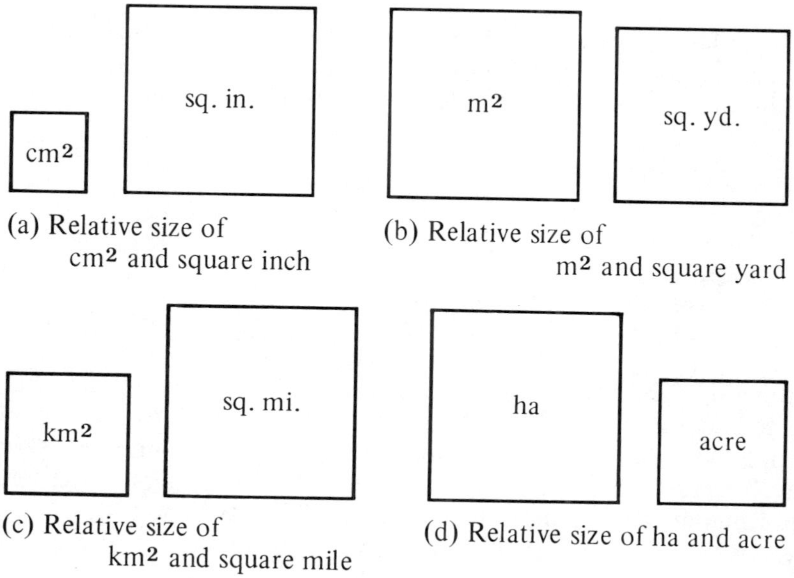

(a) Relative size of cm^2 and square inch

(b) Relative size of m^2 and square yard

(c) Relative size of km^2 and square mile

(d) Relative size of ha and acre

Figure 3.3. Relative size of pairs of units of area.

Example 1

Find the approximate number of hectares in a 160-acre farm, to the nearest hectare.

$$\begin{aligned} 1 \text{ acre} &\approx 0.4 \text{ ha, and} \\ 160 \text{ acres} &\approx 0.4\,(160) \text{ ha} \\ &\approx 64 \text{ ha.} \end{aligned}$$

Example 2

A search area of 325 km^2 includes approximately how many square miles?

$$\begin{aligned} 1 \text{ km}^2 &\approx 0.4 \text{ sq mi, so} \\ 325 \text{ km}^2 &\approx 0.4\,(325) \text{ sq mi} \\ &\approx 130 \text{ sq mi.} \end{aligned}$$

Exact Conversion Factors

Because of the fact that an inch is defined as exactly 2.54 cm, exact conversion factors for changing from English to SI units of area can be obtained.

The simplest of the exact conversion factors is found by squaring 2.54. These factors should not be memorized and are ordinarily used only when a desk calculator is available.

$$1 \text{ sq in.} = 6.4516 \text{ cm}^2$$

Two others are:

$$1 \text{ sq ft} = 929.0304 \text{ cm}^2$$
$$1 \text{ sq yd} = 0.836\ 12736 \text{ m}^2$$

As can be seen, these conversion factors, while exact, are also cumbersome and are of practical value only if a desk calculator is available. People who do need to make very precise or exact conversions usually have mechanical aids as well as reference tables. Others can generally use rougher approximations with little difficulty. The following brief exercise set emphasizes approximate conversions.

Exercise Set 7

For Exercises 1 to 12, use the approximate conversions given in the section to express each measurement correct to the nearest new unit value.

Original Measurement	*Convert Approximately to*
1. 17 cm^2	sq in.
2. 14 sq in.	cm^2
3. 35 m^2	sq yd
4. 56 sq yd	m^2
5. 214 km^2	sq mi
6. 301 sq mi	km^2
7. 230 ha	acres
8. 180 acres	ha
9. 16.52 sq in.	cm^2
10. 4,850 sq mi	km^2
11. 9.024 cm^2	sq in.
12. 0.135 km^2	sq mi

Use exact conversions for Exercises 13 to 16.

13. Convert 7 square inches to square centimeters.
14. Convert 30 square feet to square centimeters.

15. Convert 15 square yards to square meters.
16. Convert 18.2 square inches to square centimeters.

Mathematics Laboratory Activities 7

1. Combine all the approximate conversions given in the section into a single rectangular table.
2. Extend the list of approximate conversions by writing new ones, including some for very large and very small units.
3. Derive three additional exact conversion factors by using the ones given.
4. Find the price per SI unit for two items sold by area (such as dress material).
5. Find the area of your home state in square miles; then convert to square kilometers.

3.4 INTERNATIONAL UNITS OF VOLUME

The English system for measuring volume is considered confusing because of the various nondecimal equivalents and because there is one system of measurements for liquid and another for dry measures, and these systems are often poorly understood. For example, such concepts as a liquid pint or a bushel can be very fuzzy at best. Most of the difficulties of the English system are avoided in the modernized metric system.

SI Unit of Volume

The standard international unit of volume is the **cubic meter,** with the symbol m^3. Like the square meter, the cubic meter is a derived unit rather than a basic unit. Figure 3.4 represents an object $1 \text{ m} \times 1 \text{ m} \times 1 \text{ m}$ with a volume of 1 m^3.

In the modernized metric system, unlike the English system, all measures of volume are linked to the basic linear unit. To illustrate this advantage of the metric system, the reader should try to determine how such customary units as the bushel or pint are related to the inch.

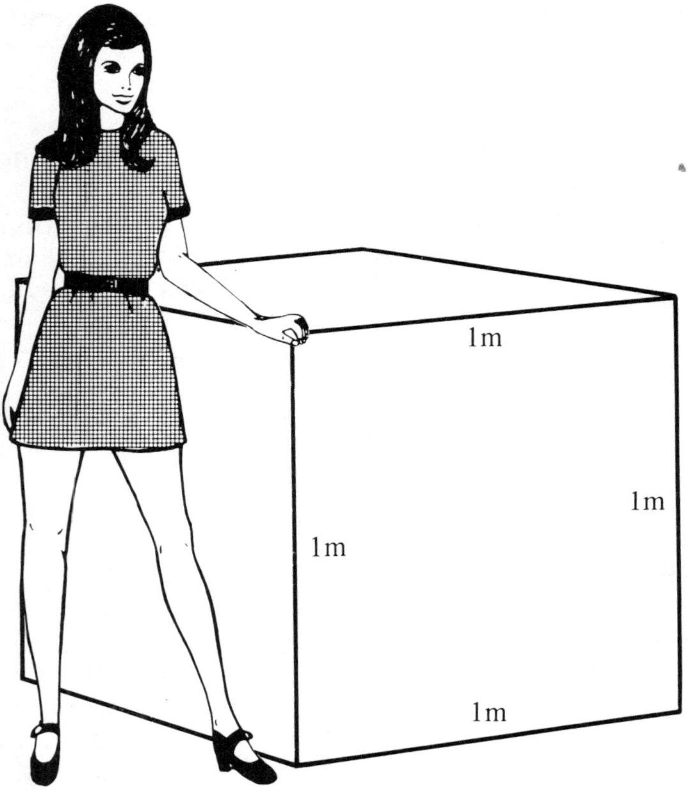

Figure 3.4. Model of 1 cubic meter.

Multiples and Submultiples

It is probable that the average person will change volume measurements only rarely and will need only the following three relationships.

$$1\ \text{dam}^3 = 1{,}000\ \text{m}^3$$
$$1\ \text{dm}^3 = 0.001\ \text{m}^3$$
$$1\ \text{cm}^3 = 0.000\,001\ \text{m}^3$$

The list of equations for multiples and submultiples of the cubic meter given below can be obtained from the list of relationships for

SI units of length, so that memorization is not necessary. For example, 1 dm = 10^{-1} m, and 1 dm^3 = $10^{-1} \times 10^{-1} \times 10^{-1}$ m^3, or 1 dm^3 = 10^{-3} m^3.

1 Mm^3	= 10^{18} m^3	1 dm^3	= 10^{-3} m^3
1 km^3	= 10^{9} m^3	1 cm^3	= 10^{-6} m^3
1 hm^3	= 10^{6} m^3	1 mm^3	= 10^{-9} m^3
1 dam^3	= 10^{3} m^3	1 μm^3	= 10^{-18} m^3

The cubic centimeter, a commonly used measure of volume, is pictured in Figure 3.5. The cubic decimeter, equal to 0.001 m^3, is also

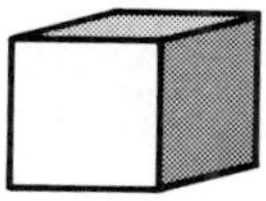

Figure 3.5. One cubic centimeter.

defined as a **liter**, with symbol l. The liter was one of the original metric units, but technically it is not an SI unit, although it is accepted by the Conference. However, it is generally used when making metric measurements of volume of liquids or capacity of containers for holding liquids. For example, in countries using the metric system, the capacity of automobile gasoline tanks is described in liters, and gasoline is sold in liters. Otherwise the expression cubic decimeter can be used. **Milliliters** (1 ml = 10^{-3} liters) are normally used in measurements of small quantities, such as medicine dosage. Note that a milliliter is equal to a cubic centimeter and is an equally common way of referring to the same quantity of liquid. The average person using the metric system does need to be familiar with both the liter and the milliliter.

Four examples illustrate change from one SI unit of volume to another.

Example 1

Express 2,639 m^3 in cubic kilometers.

$$1\ m^3 = 10^{-9}\ km^3,\ \text{and}$$
$$2{,}639\ m^3 = 10^{-9} \times 2{,}639\ km^3$$
$$= 2.639 \times 10^{-6}\ km^3.$$

Example 2

Express 0.236 km^3 in cubic meters.

$$1\ km^3 = 10^9\ m^3, \text{ so}$$
$$0.236\ km^3 = 0.236 \times 10^9\ m^3$$
$$= 2.36 \times 10^8\ m^3.$$

Example 3

Express 3.6 cm^3 in cubic meters.

$$1\ cm^3 = 10^{-6}\ m^3, \text{ so}$$
$$3.6\ cm^3 = 3.6 \times 10^{-6}\ m^3.$$

Example 4

Express 0.34 dm^3 in cubic centimeters.

$$1\ dm = 10\ cm, \text{ so}$$
$$1\ dm^3 = 10^3\ cm^3.$$
$$\text{Then } 0.34\ dm^3 = 0.34 \times 10^3\ cm^3$$
$$= 340\ cm^3.$$

Applied Problems

Applied problems with metric units of volume and capacity require appropriate operations with decimals. Three examples follow. The measures are assumed to be exact.

Example 5

Find the cost of 42.7 liters of gasoline at 12.3¢ per liter.
42.7 (0.123) = 5.2521, so the cost is $5.25, to the nearest cent.

Example 6

A container designed to hold 275 cm^3 was redesigned for a special sale to hold 12% more. How many cubic centimeters would it now hold?
275 (1.12) = 308.00, so the new container will hold 308 cm^3.

Example 7

Find the volume of a right-circular cylinder if the radius is 8.3 cm and the height is 4.7 cm. The cylinder is drawn to scale in Figure 3.6.

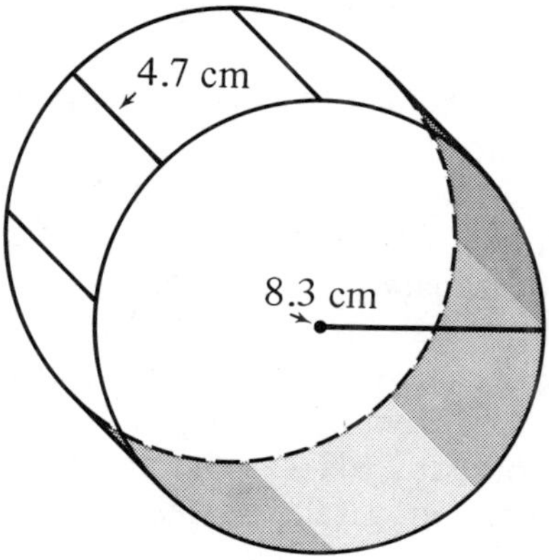

Figure 3.6. Right-circular cylinder.

The volume is $\pi(8.3)^2(4.7)$ cm^3, or 323.783π cm^3.

Exercise Set 8

For Exercises 1 to 8, change each measurement to cubic meters.

1. 9 km^3	2. 4 dm^3
3. 25 cm^3	4. 86 mm^3
5. 124 Mm^3	6. 3,418 μm^3
7. 75.4 hm^3	8. 82.9 cm^3

For Exercises 9 to 14, change each measurement in cubic meters to the unit indicated.

Original Measurement	*New Unit*
9. 5 m^3	dam^3
10. 25 m^3	hm^3
11. 128 m^3	km^3
12. 0.3 m^3	cm^3
13. 0.42 m^3	cm^3
14. 0.016 m^3	mm^3

For Exercises 15 to 18, change each measurement to the new unit indicated.

Original Measurement	*New Unit*
15. 345 km^3	hm^3
16. 0.34 cm^3	mm^3
17. 3.27 mm^3	cm^3
18. 74,826 cm^3	dam^3

19. A gasoline tank holds 60 liters. How much would it cost to fill the tank if gasoline is 12.9¢ per liter?
20. A dose of medicine is 5 cm^3. How many doses are in a bottle holding 0.7 dm^3?
21. A container holding 16.2 liters has what percent, to the nearest 0.1%, of the capacity of one holding 20.5 liters?
22. What is the capacity, to the nearest cubic meter, of a rectangular storage tank that is 3.2 m $\times$ 6.1 m $\times$ 2.9 m?
23. To the nearest cubic centimeter, what is the volume of a metal block 5.7 cm $\times$ 3.8 cm $\times$ 6.9 cm?

For Exercises 24 to 26, use 3.1416 as an approximation to π.

24. Find the volume , to the nearest cubic centimeter, of a circular cylinder if the height is 6.3 cm and the radius is 7.1 cm.
25. Find the volume, to the nearest cubic meter, of a sphere if the radius is 4.4 m.
26. Find the volume, to the nearest cubic meter, of a right-circular cone if the radius is 12.3 m and the height is 7.2 m.

Mathematics Laboratory Activities 8

1. Make up four practical problems, using metric measurements of volume.
2. Find the volume in cubic centimeters of a flashcube, a stick of butter, and a box of paper clips.
3. Find the number of cubic meters in a room.
4. Draw, with one face drawn to scale, pictures of rectangular solids with volume 1, 8, 20, 70, and 96 cm^3.
5. Find the cost per SI unit for two items sold by volume (such as gasoline).

3.5 ESTIMATES WITH SI MEASUREMENTS OF VOLUME

As with SI units of length and area, the use of standards for comparison will be helpful in acquiring the ability to estimate with SI units of volume. For example, a small wooden cube or die can be used to estimate 1 cm^3 and a large box to estimate 1 m^3. A quart carton of ice cream or a small box may help estimate 1 liter.

Several specific suggestions follow for making reasonable estimates of volume.

1. One way to estimate the volume of a small room is to think about how many large boxes 1 m^3 will completely fill it. Another way is to estimate the dimensions in meters.
2. Estimates of volume of round or irregularly shaped objects can be simplified by first thinking of a rectangular box that appears to have about the same volume.
3. Estimates of volume of smaller objects can be made by guessing how many of the objects could fit into a 1-cm^3 or 1-m^3 container.

Estimating the answers to problems with SI units of volume can be improved by practice. Here are some suggested activities. In each case, estimates should be made within the metric system; then answers should be checked in any way possible. In each case, give the answer in the appropriate metric units.

Mathematics Laboratory Activities 9

Write the estimates directly in SI units; then check by measurement or other means. Estimate the volume or capacity in SI units of a

1. room	2. large box
3. shoe box	4. dime
5. can of fruit	6. china cabinet
7. book	8. bottle
9. gas tank	10. small vial of medicine
11. ice cream carton	12. tennis ball

3.6 CONVERTING TO AND FROM SI UNITS OF VOLUME

Quick Approximate Conversion

When it is necessary to convert from or to SI units of measurement of volume, approximate conversion factors will be useful. Some of these factors are listed below:

1 qt $\approx$ 0.95 liter
1 liter $\approx$ 1.06 qt
1 cu ft $\approx$ 0.03 m^3

$1\ m^3 \approx 35$ cu ft
1 cu in. $\approx 16.4\ cm^3$

The most useful conversion from the list for the average person is from quarts to liters. In many cases, it is only necessary to remember that a liter is slightly larger than a quart. Two other common conversions are 1 liter $\approx$ 1.1 qt and 1 cu in. $\approx 16\ cm^3$.

Figure 3.7 shows the relative sizes of various SI units of volume.

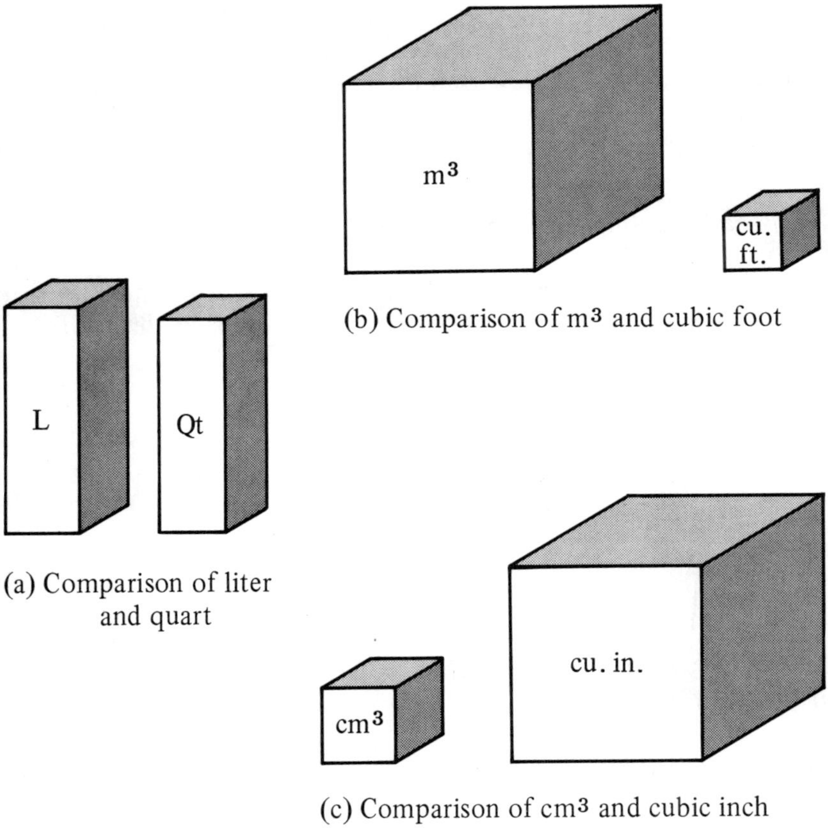

Figure 3.7. Relative sizes of SI units of volume.

The following examples illustrate the use of the approximations. Additional suggestions for practice are found in the exercises.

Example 1

Milk that sells for 29¢ per quart should sell for approximately how much per liter?
Because 1 liter is approximately 1.06 quarts, the cost should be 1.06 (29), or approximately 31¢.

Example 2

The volume of a storage tank is 75 cu ft. What is its volume, to the nearest cubic meter?
The approximate conversion factor 1 cu ft $\approx$ 0.03 m^3 can be used, so the answer is 0.03 (75) m^3, or approximately 2 m^3.

Exact Conversion Factor

The exact conversion factor from cubic inches to cubic centimeters can be found by cubing the factor 2.54 in the exact definition 1 cu in. $= (2.54)^3$ cm^3 $=$ 16.387064 cm^3, based on the corresponding definition of an inch. From this formula, exact conversion factors can be found for converting cubic feet to cubic decimeters (1 cubic foot is equal to 28.31 846 592 liters), cubic yards to cubic meters, and so on for other cubic units.

Exercise Set 9

For Exercises 1 to 10, use the approximate conversions given in the section. Give the answer correct to the nearest new unit.

1. 35 m^3 $\approx$? cu ft
2. 850 cu ft $\approx$? m^3
3. 24 cu in. $\approx$? cm^3
4. 16 liters $\approx$? qt
5. 1 gal $\approx$? liters
6. 3.2 qt $\approx$? liters
7. 42.1 liters $\approx$? qt
8. 17.2 cu ft $\approx$? m^3
9. 82.9 cu in. $\approx$? cm^3
10. 74.6 m^3 $\approx$? cu ft
11. Ice cream that sells for 89¢ per half gallon should sell for approximately how much per liter?
12. A 5-gallon gasoline can holds how many liters of gasoline?
13. Find the approximate volume, correct to the nearest cubic centimeter, of a sphere with radius 2 inches. Use 3.1416 as an approximation to π.
14. Derive the exact conversion factor for converting cubic inches to cubic decimeters.

For Exercises 15 and 16, use the fact that a liquid ounce (as a measure of capacity, not of weight) is approximately equal to 30 milliliters.

15. A bottle of medicine holding 6 ounces holds how many milliliters?
16. A bottle of medicine holding 150 ml holds how many liquid ounces?

The teaspoon (approximately 5 ml) and the tablespoon (approximately 15 ml) are measures of volume or capacity commonly used in cooking. Rewrite each measurement in Exercises 17 to 20 in milliliters.

17. 2 teaspoons
18. 3 tablespoons
19. $\frac{1}{4}$ teaspoon
20. $2\frac{1}{2}$ tablespoons

Mathematics Laboratory Activities 10

1. Combine the approximate conversions given in the section into a single rectangular table.
2. Find two other additional exact conversion factors for volume not given in the section.
3. Find, in a reference almanac, the volume in cubic meters and cubic yards of the largest dams in the world. Check some of the statements of equivalent measurements.
4. Derive additional approximate conversion factors for pairs of units not included in the list in this section.
5. Rewrite a recipe for a cake in SI units of volume or capacity, rounding to the nearest units as necessary.
6. Explain what is meant by a #2 can, using SI units.

INTERNATIONAL SYSTEM UNITS OF MASS (WEIGHT)

4

4.1 DEFINITIONS AND CONVERSION AMONG SI UNITS OF MASS

Mass and Weight

The English system of weights consists of such units as ounces, pounds, and tons. The English system of "weights and measures" does not consider mass at all. On the other hand, SI uses measurements of mass rather than weight. But SI measures of mass are set equal to units of weight. In order to understand this seemingly strange situation, it is necessary to understand the basic concepts.

In ordinary usage, weight is measured by scales, such as bathroom scales or scales at the check-out counter at the supermarket. Technically, **weight** is a measurement of the gravitational force exerted by a planet on an object. The gravitational force operates in the direction of the center of the planet. Weight varies according to (1) the distance from the object to the center of the planet, (2) the mass of the planet itself, and (3) the mass of the object. This information is summarized pictorially in Figure 4.1. As is commonly known, the weight of a man is different on the moon than on the surface of the earth. Because of the fact that the weight of an object changes according to its position, weight is not an invariant property that can be measured once and for all. As a result it is not really appropriate for use in a scientific system of measurement such as SI.

Rather than measure weight, which is a varying quantity for a

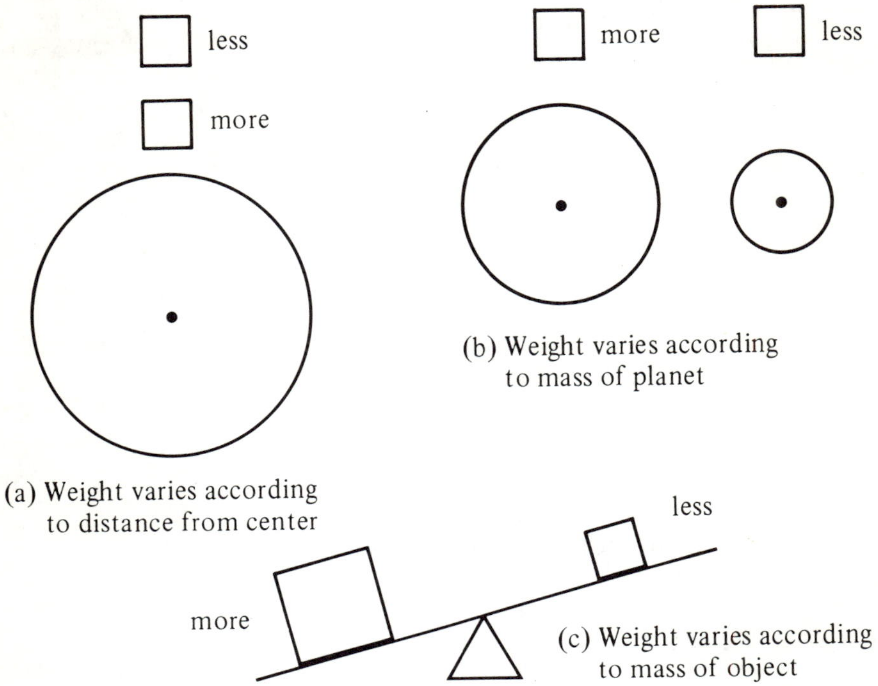

Figure 4.1. Variations in weight.

particular object, scientists using the International System prefer to measure **mass**. The word "mass" has already been used without definition in the explanation of the concept of weight. Probably the simplest intuitive notion of what is meant by mass is that it measures the amount of matter in an object. For example, a big block of wood contains more mass than a smaller block of wood of the same kind. A chunk of iron has more mass than a feather.

From a scientific point of view, however, an explanation of mass as amount of matter is not sufficient. Instead, mass is defined as a measure of the **inertia** of an object. Intuitively, inertia means resistance to change. More specifically, inertia means resistance to being speeded up (**acceleration**) or slowed down (**deceleration**). It would seem reasonable that it would take more force to stop a block of iron moving through space than it would to stop a feather moving through space. It would seem reasonable that it would take more force to stop the movement of a large man than a small boy.

The concepts of force, mass, and acceleration are related by a physical law called **Newton's Second Law of Motion**. This law states that $F = ma$, where F is force, m is mass, and a is acceleration. In physics, **force** affects motion and is measured by the change in velocity that it produces on a mass in a certain time. The SI unit of force, a derived unit, is the **newton**, named after Sir Isaac Newton. One newton is the force needed to cause an increase in velocity of one meter per second, for a one-kilogram mass.

Basic SI Unit of Mass

In the International System of measurements, the basic unit for mass is called the **kilogram,** with symbol kg. Unlike other metric units today, the kilogram is still defined in terms of a **prototype,** an arbitrary cylinder of platinum-iridium alloy from which duplicates are made to preserve in various national bureaus of standards. The kilogram was originally intended to be the mass of one cubic decimeter (a liter) of water at the temperature of maximum density, but later it was found that the kilogram actually corresponded to 1.000028 dm^3 of water. As a result of an error in the measurement, the prototype definition was used instead.

Although ordinary scales measure weight, they cannot actually measure mass. It should occur to the reader that a method is needed for comparing the mass of an object with that of the prototype. Fortunately, this step can be accomplished by means of an equal-arm balance as pictured in Figure 4.2. Actually, the balance compares weights rather than masses. But with the assumption that the same

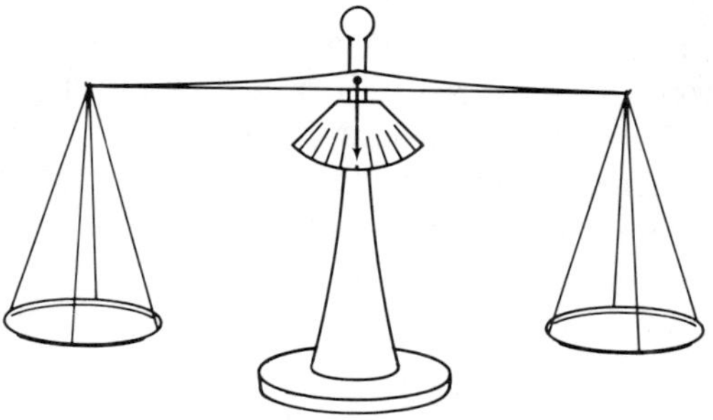

Figure 4.2. Equal-arm balance.

force of gravity is acting on both pans of the balance, the result is an indirect comparison of the two masses. This helps to explain why equations are written connecting units of mass and weight. If two objects have the same mass, then they will have the same weight at the same place and under the same circumstances.

Multiples and Submultiples

The most important comparisons of SI units of mass are 1 metric ton = 1,000 kg and 1 kg = 1,000 grams. The relationships between some of the most common metric units of mass are given below.

1 metric ton (t)	$= 10^3$ kg	$= 10^6$ g
1 kilogram	$= 10^3$ g	
1 milligram	$= 10^{-3}$ g	
1 microgram	$= 10^{-6}$ g	

Recall from Chapter 1 that the original metric unit of mass was the gram, not the kilogram. The result has been a certain deviation from the standard SI way of naming the units as powers of 10 times the basic unit. Even though the kilogram is the basic unit for which the prototype was constructed, prefixes based on the gram were retained. Consequently, the same prefixes as for length can be used to define other units of mass between and beyond those given.

There are at least two other deviations from the standard names, however, in addition to the fact that they are not based on the fundamental unit. First, one thousand kilograms is normally called a metric ton (t) rather than the megagram, as might be expected. Second, the carat, adopted in 1907, is a unit used in conjunction with SI that is not a power of 10 times a basic unit. Normally used to measure diamonds and precious stones, but technically not an SI unit, the carat is defined as $\frac{1}{5}$ gram, or 200 milligrams.

Two examples of the use of metric relationships of mass follow:

Example 1

Express 0.453 t in kilograms.

$$1 \text{ t} = 1{,}000 \text{ kg, and}$$
$$0.453 \text{ t} = 1{,}000\,(0.453) \text{ kg}$$
$$= 453 \text{ kg.}$$

Example 2

Express 792 g in kilograms.

$$
\begin{aligned}
1\ \text{g} &= 10^{-3}\ \text{kg, and} \\
792\ \text{g} &= 792\,(10^{-3})\ \text{kg} \\
&= 0.792\ \text{kg.}
\end{aligned}
$$

The answers contain the same sequences of digits, with the decimal points appearing in different positions as expected, so that the examples could be worked orally.

Computation with SI Units of Mass

Three examples are given to illustrate computation with SI measurements of mass. The measures are considered to be exact.

Example 3

Find the combined mass of two objects if one has a mass of 3.54 kg and the other has a mass of 6.48 kg.
The combined mass is 10.02 kg because 3.54 + 6.48 is 10.02.

Example 4

A mass of 8.643 g is what percent of a mass of 9.757 g?
$8.643 \div 9.757 \approx 0.886$, so the answer is approximately 88.6%.

Example 5

What is the approximate total worth of a metric ton of dried fruit valued at 89¢ per 453-gram box?
The value may be written as approximately $0.001965 per gram, so that the value for 1 ton (10^6 g) is $0.001965 ($10^6$), or approximately $1,965. This answer may be compared with the answer obtained by first finding the cost for 453 metric tons and then dividing by 453.

Derived Units

In the modernized metric system, many other derived units of measurement are obtained from the basic units by definition. For example, the square meter for area and the cubic meter for volume are derived from the basic unit, the meter, for length. Derived from the basic unit of mass are the newton, a unit of force defined previously, the **pascal**, a unit for pressure, the **joule**, a unit for work or energy, and the **watt**, a unit for power. One pascal is a pressure of one newton per square meter. A joule is the amount of energy required to push a distance of one meter against a force of one newton. A watt is one joule of energy per second. With the exception of the watt, the units derived from mass are rarely encountered by the average person using SI units along with elementary mathematics. All are important in scientific applications, however. The concept of derived units is discussed further in Chapter 5.

Exercise Set 10

For Exercises 1 to 10, express each mass in the units indicated.

Original Mass	*New Unit*
1. 34 kg	g
2. 5 t	kg
3. 6 g	mg
4. 156 μg	g
5. 350 mg	g
6. 7.52 kg	g
7. 942.7 g	kg
8. 0.314 g	mg
9. 7.42 mg	μg
10. 63.2 μg	g

In Exercises 11 to 14, express each mass as a fraction of a carat.

11. 50 mg
12. 100 mg
13. 40 mg
14. 10 mg
15. A new truck could haul 14% more than an old truck, which could haul 2 t. How much could the new truck haul?
16. A metric ton of dried fruit could be packed into how many boxes containing 500 g each?
17. Atomic explosions are often measured in megatons. A one-megaton explosion releases as much energy as one million metric tons of TNT. Express one megaton in terms of kilograms.

For Exercises 18 to 19, the numbers refer to micrograms of pollutants per cubic meter, a common way of measuring air pollution.

18. A city whose mean measurement of nitrogen dioxide was 275 adopted a plan to reduce the amount by 25% during the next year. If the plan was successful, what would be the new mean measurement, to the nearest whole number?
19. Two cities had mean measurements of sulfur dioxide of 84 and 69. Express the difference in grams per cubic kilometer.
20. The density of an object is a measure of its mass per unit of volume. The density of the sun is approximately 1.4 g/cm^3, and the density of the earth is approximately 5.52 g/cm^3. The density of the sun is about what percent of the density of the earth?
21. A mass of one kilogram of seawater contains about 18,980 mg of chlorine, 10,561 mg of sodium, and 1,272 mg of magnesium, as well as smaller amounts of other elements. The combined mass of these three elements is what percent, to the nearest 0.1%, of the total mass of the water?
22. One serving of a certain kind of cereal contains 33% of the thiamin, riboflavin, and iron and 6% of the calcium recommended as minimum daily adult requirements. If the requirements are thiamin, 1.0 mg, riboflavin, 1.6 mg, iron, 10 mg, and calcium, 0.8 g, find the amount of each in one serving of the cereal.
23. If the contents are the same quality, which would be a better buy, a 650-g box at 79¢ or a 1,100-g box at $1.40?
24. It takes 15 kg of a certain kind of fertilizer for a 1,000-m^2 lawn. This is an application at the rate of how many grams per square meter?

Mathematics Laboratory Activities 11

1. Make up four practical problems involving metric measurements of mass.
2. Find the mass in SI units of five common objects.
3. Find your weight in kilograms.
4. Check tables of recommended weights in kilograms for children of various ages.
5. In a reference text find the mass and density of each planet in SI units.
6. Construct models of wood or metal whose mass is approximately 1 gram and 1 kg.

7. Find three particular objects (possibly shoes, a tool, or a book) that have a mass of approximately 1 kg each.

4.2 ESTIMATING WITH SI UNITS OF MASS

The ability to estimate with SI units of mass can be improved by making careful guesses and then making the actual measurements. Some standards to use in helping with estimates might include your own weight in kilograms, the weight of a small block of wood, and the fact that an ordinary brick has a mass of approximately 2 kg. It may be helpful to know that a paper clip has a mass of about 1 gram and that a flashcube has a mass of about 10 grams. Make a list of reference weights or masses before continuing.

Several approaches to estimating mass are as follows:

1. To estimate the mass of a pen, try to see approximately how many paper clips might have the same mass.
2. To estimate the mass of a child, think of it as a decimal part of your own mass.
3. To estimate the mass of a large truck, compare it with a smaller truck with a mass of one metric ton.

Mathematics Laboratory Activities 12

Estimate the mass of each object in SI units; then check.

1. a particular child
2. bag of apples
3. glass of water
4. bag of potatoes
5. cube of sugar
6. postage stamp
7. large loaded truck
8. large jet airplane
9. wristwatch
10. carton of milk
11. loaf of bread
12. level teaspoon of sugar
13. dozen eggs
14. gallon of water

4.3 CONVERTING TO AND FROM SI UNITS OF MASS (WEIGHT)

Approximate Conversion

Metric units of mass are commonly set equal to ordinary units of weight, as discussed in Section 4.1. The weights used in this chapter are those called **avoirdupois** weights. In addition, there is a system of **troy** weights for precious metals and **apothecaries'** weights for medicines. The ton used here is 2,000 pounds, so that the complication of the long and short ton is not introduced. Here is a useful table of rough approximations:

1 oz ≈ 28 g	1 g ≈ 0.035 oz
1 lb ≈ 0.45 kg	1 kg ≈ 2.2 lb (or 2lb 3 oz)
1 ton ≈ 0.9 t	1 t ≈ 1.1 ton (or 1 ton 200 lb)

Figure 4.3 compares various units of mass or weight. Conversion factors are used in the following three sample applications.

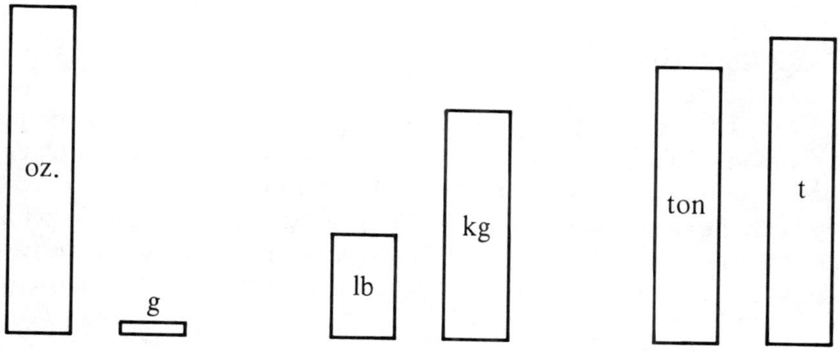

(a) Comparison of ounce and gram

(b) Comparison of pound and kilogram

(c) Comparison of ton and metric ton

Figure 4.3. Comparisons of units of weight or mass.

Example 1

A processor of fruit needs to label a can with the metric units of weight and the approximate English equivalent. If the weight is 582 g, what is the approximate weight in ounces, to the nearest ounce?

$$1 \text{ g} \approx 0.035 \text{ oz, so}$$
$$582 \text{ g} \approx 582\,(0.035) \text{ oz}$$
$$\approx 20 \text{ oz.}$$

An alternative method of solution if the weight is desired in pounds comes from the fact that 450 g ≈ 1 lb, so 582 g ≈ 582/450 lb, or ≈ 1.3 lb.

Example 2

What is the approximate weight in kilograms, to the nearest kilogram, of a person who weighs 137 lb?

Since 1 lb $\approx$ 0.45 kg,

$$137 \text{ lb} \approx 0.45\,(137) \text{ kg}$$
$$\approx 62 \text{ kg}.$$

Example 3

If meat sells for \$1.29 per pound, find the approximate cost per kilogram.

The cost should be approximately \$1.29 (2.2) or about \$2.84 per kilogram.

Exact Conversion

The exact conversion from pounds to kilograms is made possible because the pound has now been defined as equal to 0.453 592 37 kg. Based on this exact conversion factor, for example, is the exact conversion statement 1 oz = 28.349 523 125 g. Obviously, approximations are usually sufficient for everyday purposes, since the exact conversion factors are cumbersome and should be used only with a desk calculator.

Exercise Set 11

Use the approximate conversion factors to convert to the nearest new unit indicated, for Exercises 1 to 12.

Original Measurement	*New Unit*
1. 12 oz	g
2. 16 g	oz
3. 150 lb	kg
4. 130 kg	lb
5. 12 tons	t
6. 35 t	ton
7. 3.72 oz	g
8. 5.4 lb	kg
9. 63.2 g	lb
10. 3.5 kg	oz
11. 4.31 t	ton
12. 6.24 tons	t

13. What is the approximate weight in grams of a box to be mailed that weighs 1 lb 5 oz?
14. The manufacturing plant consumed 120,000 tons of materials. Approximately how many metric tons was this?
15. What is the approximate weight in grams of an 8-oz glass of milk?
16. A box of gelatine weighing 85 g weighs how many ounces, to the nearest ounce?
17. If apples sell for 29¢ per pound, what should be the corresponding cost per kilogram?
18. If candy sells for 10¢ per ounce, what should be the corresponding cost per gram?
19. If a 10-lb bag of potatoes sells for 79¢, what is the corresponding cost per kilogram?
20. If a package of meat selling for $4.23 weighed 2 lb 7 oz, find the approximate cost per kilogram.
21. Use approximate conversions to give the weight, to the nearest 0.1 kg, for a 7-lb 9-oz baby.
22. A company sold fertilizer that required 32 lb for 10,000 square feet. Rewrite the information approximately, using kilograms and square meters.

For Exercises 23 and 24, use exact conversion factors and then round to two significant digits.

23. Convert 2.5 lb to kilograms.
24. Convert 1.7 oz to grams.

Mathematics Laboratory Activities 13

1. Combine the approximate conversions given in the section into a single rectangular table.
2. Find one additional exact conversion factor by using those given.
3. Use a table of conversion factors and a desk calculator to practice making very precise changes from one system of weight measurement to another.
4. Find examples of products in grocery stores marked in both SI and English units. Check to see what conversion factors were used.
5. Find the price per SI unit for three items sold by weight (such as sugar).

ADDITIONAL INTERNATIONAL SYSTEM UNITS

5

5.1 INTERNATIONAL SYSTEM MEASUREMENT OF TEMPERATURE

Basic Unit for Temperature

The remaining basic SI unit of greatest interest and usefulness to nonscientific users of mathematics is the basic unit for temperature. The basic unit for measuring temperature in the modernized metric system is the **kelvin** (abbreviated K), named in honor of Lord Kelvin (William Thompson, 1824–1907). The kelvin is defined as the fraction 1/273.16 of the thermodynamic temperature of the triple point of water. This temperature is measured by using a special cell that is a glass cylinder containing pure water. The cell is then cooled to a certain point, which is designated as 273.16 K. This is the **triple point**, the temperature and pressure at which solid, liquid, and vapor states of water are in equilibrium. The triple point, slightly above freezing, is the temperature in which water can exist in three states.

The Celsius Scale

Although the kelvin is the basic unit of temperature in the International System, temperature is measured on what is known as the **Celsius** scale (abbreviated C), originally referred to as the centigrade scale. In 1948 the ninth General Conference on Weights and Measures adopted the name Celsius in honor of Anders Celsius,

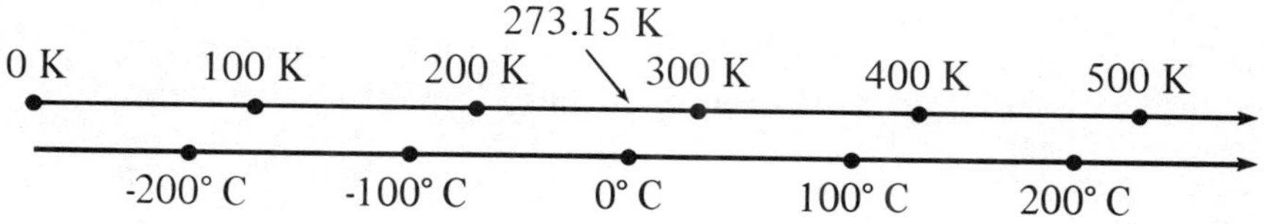

Figure 5.1. Kelvin and Celsius scales.

a Swedish scientist (1701–1744). Figure 5.1 relates the Kelvin and the Celsius scales. Zero Kelvin is defined as absolute zero, while 0° Celsius is taken as 273.15 K. Absolute zero is the lowest theoretical temperature possible for a gas. Then one Celsius degree is defined as one kelvin degree. On the Celsius scale, water freezes at approximately 0°C and boils at approximately 100°C. The use of 100 units from freezing to boiling makes the Celsius scale convenient to use with decimal notation. Here is another important instance of the superiority of the metric system and its connection with standards from nature.

Two examples help to explain the relationship between the absolute Kelvin scale and the Celsius scale.

Example 1

Express 18°C in kelvins.
The answer is (273.15 + 18) K, or 291.15 K.

Example 2

Express 300 K on the Celsius scale.
The answer is (300 − 273.15)°C, or 26.85°C.

Estimating Temperature

In order to learn how to estimate by using the Celsius scale, it is important to acquire an actual thermometer to check the results and to try laboratory activities, such as those at the end of the section, in order to improve your estimating. The use of the freezing and boiling point on the Celsius scale, the Celsius reading for normal body temperature, and experiences with climate and temperature readings to know what is cold, hot, or moderate in Celsius units will make estimating much easier. A particular way to develop a sense of the Celsius scale in terms of hot and cold is to start with 0°C, 5°C, 10°C, and so on, deciding if a person on a street at that temperature would be cold, comfortable, or hot. For example, temperatures of 15 to 25°C might be considered comfortable, but 30°C might be too warm.

Conversion

The exact conversion back and forth from the Celsius scale to the Fahrenheit scale depends on the formula $F = \frac{9}{5}C + 32$. This formula can be derived from the knowledge that a linear equation connects the two temperature units and from the information that if $C = 0, F = 32$ and if $C = 100, F = 212$. (See Figure 5.2.) Let the relationship be of the form $F = aC + b$, with a and b to be determined. Since $C = 0$ when $F = 32$, we have $32 = 0 + b$, or $b = 32$. This means that the relationship is of the form $F = aC + 32$. If

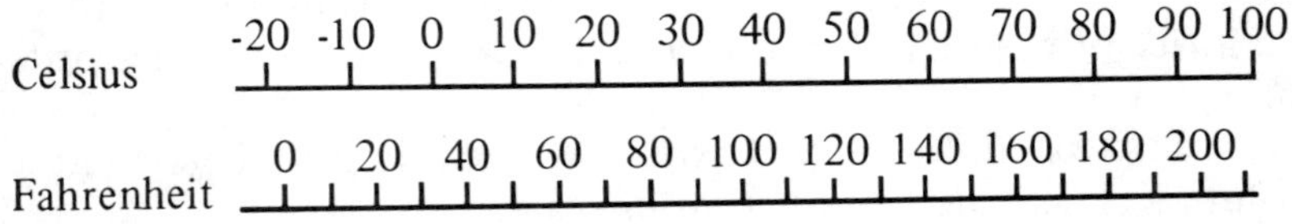

Figure 5.2. Celsius and Fahrenheit scales.

$C = 100$, then $F = 212$, so that

$$\begin{aligned} 212 &= a(100) + 32, \\ 180 &= 100a \\ \tfrac{9}{5} &= a, \text{ and thus } F = \tfrac{9}{5}C + 32. \end{aligned}$$

This formula can be solved for C as a function of F. $C = \frac{5}{9}(F - 32)$. But one formula suffices for conversions either way. Examples illustrate the conversion both ways, using the first formula given.

Example 3

Convert 80°F to the Celsius scale, correct to the nearest unit.
$80 = \frac{9}{5}C + 32.$
$48 = \frac{9}{5}C$
$27 \approx C.$
The Celsius temperature is approximately 27°C.

Example 4

Convert 12°C to the Fahrenheit scale, correct to the nearest unit.

$$\begin{aligned} F &= \tfrac{9}{5}(12) + 32 \\ &= \tfrac{108}{5} + 32 \\ &= 21\tfrac{3}{5} + 32 \\ &\approx 54. \end{aligned}$$

The Fahrenheit temperature is approximately 54°F.

One useful technique, in both approximate and exact uses of the formula $F = \frac{9}{5}C + 32$, is based on the fact that one solution with whole numbers is 20°C, 68°F. The equation 20°C = 68°F can be used as a starting point. Then, utilizing the knowledge that each five Celsius degrees is the same as nine Fahrenheit degrees, additional equivalent statements may be written easily. For example, 25°C = 77°F and 15°C = 59°F.

Exercise Set 12

For Exercises 1 to 4, convert to the Celsius scale.

1. 350 K
2. 45 K
3. 63.5 K
4. 451.7 K

For Exercises 5 to 10, express the measurements in the Fahrenheit scale, correct to the nearest unit.

5. 18°C
6. 26°C
7. − 7°C
8. − 4°C
9. 26.4°C
10. 21.7°C

For Exercises 11 to 16, express the answer in the Celsius scale, correct to the nearest unit.

11. 70°F
12. 90°F
13. 20°F
14. − 7°F
15. 350°F
16. 98.4°F

For Exercises 17 to 20, write equivalent statements derived from 20°C = 68°F to find the Fahrenheit equivalent of the Celsius temperature given.

17. 30°C
18. 35°C
19. 10°C
20. 5°C
21. A cookbook specifies that bread should be baked at 375°F. If the book is rewritten using degrees Celsius, what should be the temperature on the Celsius scale, to the nearest ten degrees Celsius?

For Exercises 22 and 23, express each normal temperature in the Celsius scale, correct to the nearest 0.1 unit.

22. 98.6°F (humans)
23. 101.5°F (dogs)

24. A child with a fever had a temperature of 102.1°F. What was the temperature on the Celsius scale, to the nearest 0.1 unit?

For Exercises 25 and 26, use the fact that one calorie is defined as the amount of heat required to raise the temperature of one gram of water through one Celsius degree. (As used in common language to refer to food, the kilocalorie is the actual unit used.)

25. How many calories are required to raise the temperature of 12 g of water from 5°C to 13°C?
26. How many calories are required to raise the temperature of 18.3 g of water from 8.1°C to 16.7°C?

Mathematics Laboratory Activities 14

For Activities 1 to 8, estimate the answer directly in Celsius degrees and then check by measuring or using reference sources.

1. What is the present temperature in Celsius units?
2. About how hot would it be in the middle of a desert in the middle of summer?
3. At what temperature should a room be kept in order for it to be comfortable?
4. What is the approximate normal operating temperature of an automobile engine?
5. What is the average low temperature during January in Bismarck, North Dakota?
6. At what temperature should roast beef be cooked in an oven?
7. At what temperature does lead melt?
8. What is the average high temperature during January in Miami, Florida?
9. Study weather records for various cities giving average temperatures each month in Celsius degrees.
10. Prepare a chart showing corresponding Fahrenheit and Celsius readings for various Fahrenheit readings from 0 to 100°F.

5.2 SI UNITS FOR TIME, CURRENT, LUMINOUS INTENSITY, AND AMOUNT OF SUBSTANCE

Base units in the International System for measuring length, mass, and temperature have been discussed in previous sections. These units are frequently encountered in mathematical applications of measurement.

The four additional base units in the International System are the **second** for measuring time, the **ampere** for measuring electrical current, the **mole** for measuring the amount of an element or compound in chemistry, and the **candela** for measuring luminous intensity. In none of these cases, however, is there really a corresponding system of units in the English system, so conversion is unnecessary.

Although the average person should be aware of the ampere, mole, and candela as basic SI units, he will not need to use these units for computation. They are included in this text so that it can serve as a more useful reference when needed by a person encountering the SI units in scientific applications.

SI Measurement of Time

Modern scientists realized that the second defined as part of a mean solar day had become unsatisfactory because of variations in the speed of rotation of the earth. Instead, as for most other units in the International System, the present definition involves concepts from atomic physics. In 1964 the General Conference defined an atomic second, using periods of radiation of cesium-133. The atomic definition became the sole definition in 1967.

Cesium-133 atoms are made to pass through a system of magnets and a resonant cavity into a detector. An oscillator is timed to the frequency of the atoms. When 9,192,631,770 oscillations have occurred, the **atomic clock** indicates that one second has elapsed. In the SI plan for measuring time, prefixes particularly recommended are ks, ms, μs, and ns, although minute, hour, and day are still included as commonly used units. The nanosecond is frequently used in circuit design for computers.

The standard unit for velocity in the metric system is **meters per second** (abbreviated m/s). Two examples are given to show conversion within the metric system and also from the English system.

Example 1

Convert 100 km/hr to meters per second, approximately.
100 km/hr is 100,000 m/hr, or approximately
28 m/s.

Example 2

Convert 12 yd/s to meters per second, to the nearest unit.

$$\begin{aligned} \text{Since } 1 \text{ yd} &\approx 0.9 \text{ m}, \\ 12 \text{ yd/s} &\approx 12\,(0.9) \text{ m/s} \\ &\approx 11 \text{ m/s}. \end{aligned}$$

SI Measure of Electrical Current

The ampere, an International System unit, is probably familiar as a measure of the rate of flow of electric current, but its definition is given to show how various units in the metric system are related. Suppose that two parallel wires are one meter apart, as shown in Figure 5.3. One **ampere** (A) is the amount of current that must be

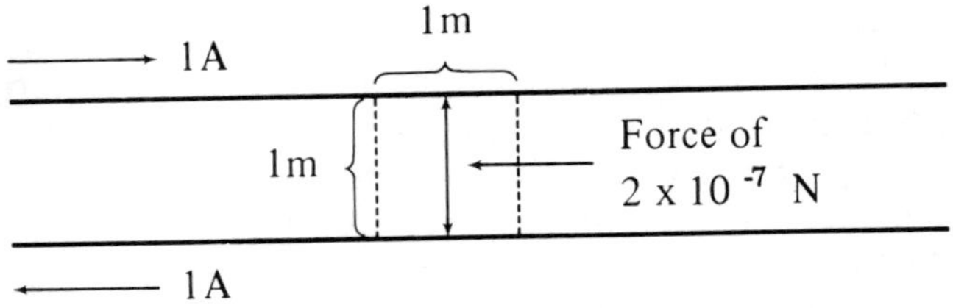

Figure 5.3. Diagram for definition of ampere.

maintained in the wires to produce a force between them of 2×10^{-7} newtons per meter of length of the wires. Recall that the newton is the international unit of force.

One ampere of current is approximately the amount required for a 100-watt light bulb. The small amount required for some scientific instruments is measured in **microamperes,** whereas factories may measure needs in terms of **kiloamperes.** Amperes are measured by using a relatively common instrument called an **ammeter**, shown in Figure 5.4.

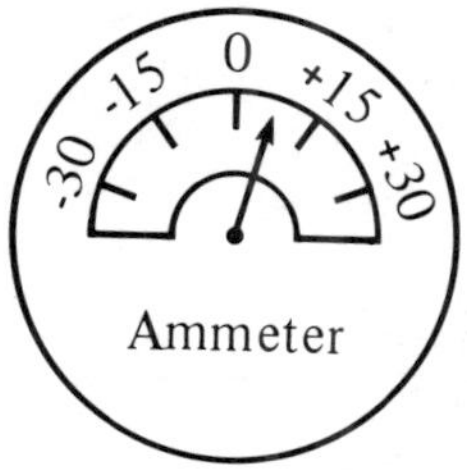

Figure 5.4. Ammeter.

Example 3

Express 2,314 μA in kiloamperes.

$$1\ \mu A = 10^{-6}\ A$$
$$= 10^{-9}\ kA,$$
$$\text{So } 2{,}314\ \mu A = 2{,}314\ (10^{-9})\ kA, \text{ or}$$
$$= 2.314 \times 10^{-6}\ kA.$$

Other SI-derived units associated with current are the **volt** and the **ohm**. International electrical units based on the metric system have been the legal units of electricity measurement in the United States since 1894.

SI Measure of Luminous Intensity

The SI unit for luminous intensity is the **candela.** The metric units of area and temperature are used in defining the candela. It is the luminous intensity of 1/600,000 of a square meter of a black body at a temperature of 2045 K (the temperature of freezing platinum). In theoretical physics a **black body** is a perfect absorber and emitter of radiation. As actually used, a black body is a device similar to the inside of a camera, a hollow chamber provided with heating appliances and with an opening for the emission of the internal radiation.

The candela is a modern unit for measuring the amount of light produced by a light source, replacing the earlier unit called **candlepower,** which is the amount of light given off by a candle of a standard size. The candela is slightly less than one candle power.

A light source of 1 candela whose light is evenly distributed in each direction radiates a **light flux** of 4π lumens. The **lumen** is an SI-derived unit. The lumen can be regarded as a measure of how bright

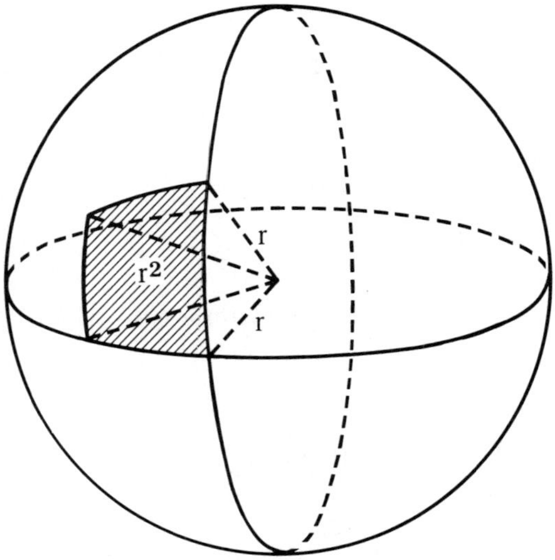

Figure 5.5. Unit square of surface illuminated at 1 lumen by source of 1 candela.

the light is when it strikes a surface. The definition can be explained in Figure 5.5. A source of 1-candela intensity at the center of a unit sphere will illuminate the inside surface of the sphere (with surface area 4π) at a total of 4π lumens. Each unit square of the surface is illuminated at 1 lumen. Light-bulb cartons ordinarily give both the number of watts and the average output in lumens. For example, a 60-watt bulb has an output of about 870 lumens, and a 100-watt bulb has an output of about 1,750 lumens.

SI Measure of Amount of Substance

The seventh basic SI unit was added only recently (in 1971). The **mole** is the unit that describes the quantity of an element or compound. The SI unit of mass occurs in its definition. This unit is the amount of substance that contains as many elementary particles as the number of atoms in 0.012 kg of carbon-12. The elementary particles are specified as atoms, molecules, electrons, and so on. The number of molecules contained in 1 mole of a substance is a number called **Avagadro's constant**, equal to 6.0225×10^{23}. The mole is useful in chemistry for measuring the amounts of chemicals involved in reactions. Derived from the mole is an SI unit of concentration of amount of substance called the **mole per cubic meter.** Moles per liter is also frequently used in chemistry texts.

Exercise Set 13

For Exercises 1 to 8, convert each measurement to meters per second. For Exercises 5 to 8, use approximate conversion factors.

1. 35 km/s
2. 6 dam/s
3. 15 cm/s
4. 250 mm/s
5. 5 yd/sec
6. 2 miles/sec
7. 13 ft/sec
8. 4 in./sec

For Exercises 9 to 12, use the fact that the velocity of light is 299,792.5 km/s.

9. What is the velocity of light in meters per second?
10. How far does light travel in a kilosecond?
11. How far does light travel in a millisecond?
12. How far does light travel in a microsecond?

For Exercises 13 to 16, convert each measurement to the unit indicated.

13. 3.4 kA to A
14. 5.6 A to μA
15. 382 A to kA
16. 453.7 μA to A

For Exercises 17 and 18, what is the light flux in lumens radiated by a source of this intensity?

17. 5 candela
18. 30 candela

Mathematics Laboratory Activities 15

1. Compare tables showing differences between atomic and solar time measurements.
2. Find practical applications where the velocity of various objects is known. Convert each to meters per second if it is given in other units.
3. Find the average number of lumens for other sizes of light bulbs besides those given.
4. Find, in a chemistry reference text, the mole per cubic meter or mole per liter for various common substances.

5.3 SUPPLEMENTAL SI UNITS OF ANGLE MEASURE

The Radian

The seven basic SI units (meter, kilogram, kelvin, second, ampere, candela, mole) have been discussed previously. Two geometric units of angle measure have been incorporated as supplemental units for the modernized metric system. The first of these, the **radian,** is familiar to anyone who has studied trigonometry. A central angle of a circle of radius *r* has a measurement of 1 radian if it is subtended by an arc equal to *r*. Thus, in Figure 5.6, if arc *AB* is the same length as the radius, the central angle has a measurement of 1 radian.

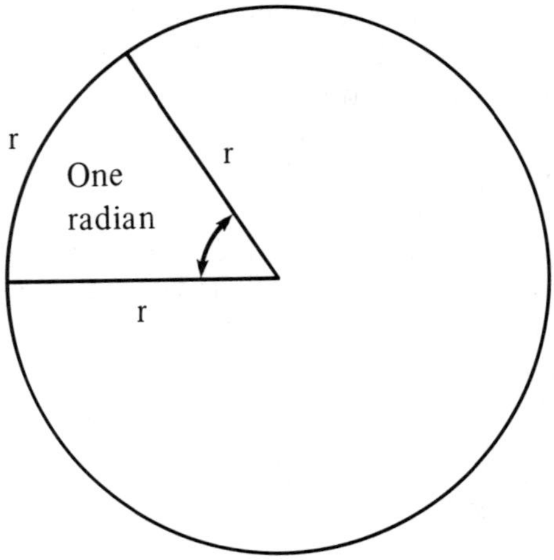

Figure 5.6. Concept of radian

A central angle thought of as subtended by the entire circle would have a radian measure of 2π. A right angle is subtended by one-fourth the circle, so its radian measure is $\pi/2$. The sum of the measures of the angles of a triangle is π radians.

The radian measure for angles is the ordinary one used in mathematics; hence it is not necessarily related only to SI units. The radian measure for angles is simple to use in a decimal system. Also, on a graph showing trigonometric functions of angles, for example, the same scale can be used for both the horizontal and the vertical axes if radian measures are used.

An example illustrates computation with radian measure. In abstract mathematical problems, the exact value π is ordinarily retained in the answer, but in applied problems with measurement, decimal approximations are common. Here 3.1416 is used as an approximation to π.

Example 1

> If two angles of a triangle have measures in radians of 1.1204 and 1.7986, find the measure of the third angle.
> The third angle has an approximate radian measure of $3.1416 - (1.1204 + 1.7986) = 0.2226$.

Estimating in radians can be improved by trying various-sized angles as suggested in the laboratory activities at the end of the section and then measuring with a protractor marked in radians. A further aid is to remember the decimal approximation 1.57 for a right angle and 0.785 radian for half a right angle.

The conversion to and from radians can be accomplished by the use of readily available tables giving the new measurement to a certain number of significant figures. A less-precise conversion comes from the fact that 1 radian is approximately equal to 57°. The exact conversion is not in terms of finite decimals but from the definition of the radian.

The exact statement is 2π radians $= 360°$, from which may be derived such statements as $60° = \pi/3$ radians and $45° = \pi/4$ radians.

Example 2

> Express 2 radians approximately and exactly in degrees.
> Since 1 rad $\approx 57°$,
> 2 rad $\approx 114°$.
> Since 2π rad $= 360°$,
> 2 rad $= (360/\pi)°$.

The Steradian

The second supplementary unit for the metric system is the three-dimensional analog of the radian, called the **steradian.** A

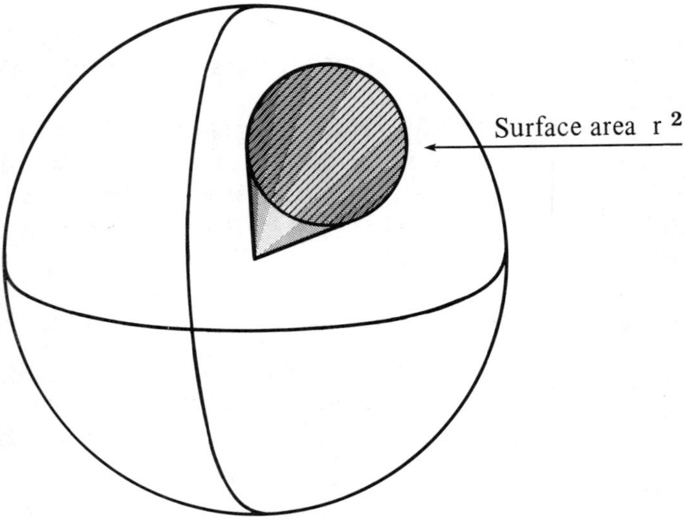

Figure 5.7. Concept of steradian.

steradian is a solid angle with its vertex at the center of a sphere and subtended by a spherical surface area r^2, for r the radius of the sphere. (See Figure 5.7.) Since the entire surface area is $4\pi r^2$, a steradian is subtended by about $\frac{1}{12}$ of the entire surface area of the sphere. A solid angle subtended by the entire sphere is equal to 4π steradians. The concept of steradian is related to that of a lumen discussed in Section 5.2.

Exercise Set 14

For Exercises 1 to 4, give the exact measurement in radians.

1. $\frac{1}{2}$ of a right angle
2. $\frac{1}{3}$ of a right angle
3. $\frac{3}{4}$ of a right angle
4. $\frac{3}{2}$ of a right angle

For Exercises 5 to 8, use 3.1416 for π. Find the approximate radian measure of the supplement of each angle. The measure is given in radians.

5. 2.1423
6. 1.7964
7. 0.0125
8. 0.0042

For Exercises 9 and 10, if two angles of a triangle have the measurements given in radians, find the approximate measurement of the third angle.

9. 0.7341 and 0.6892
10. 1.0028 and 1.0257

For Exercises 11 to 16, convert the measurement exactly to radians.

11. 30°
12. 60°
13. 120°
14. 15°
15. 270°
16. 10°

For Exercises 17 to 20, express the measurement exactly in degrees.

17. $\frac{5\pi}{4}$
18. $\frac{\pi}{8}$
19. $\frac{5\pi}{8}$
20. $\frac{\pi}{12}$

For Exercises 21 to 24, find the measurement in steradians of an angle subtended by the following portions of the entire sphere.

21. one-half
22. one-fourth
23. 0.1
24. 0.01

Mathematics Laboratory Activities 16

1. Draw representations of five different angles. Estimate the radian measure of each; then check by measurement.
2. Draw a triangle. Find the approximate radian measurement of each angle. Compare the sum of the three measurements with 3.1416.
3. Describe an instrument that you think could be built for measuring angles in steradians. Build a simple model if possible, or make a sketch.

5.4 A COHERENT SYSTEM OF UNITS

Although each SI basic unit has been discussed separately, it should be obvious that the units are interrelated. For example, the concept of meter appears in the definitions of ampere and candela. Associated with the seven basic SI units are a great many other units called derived units, some of which have been included in previous discussions. Each derived unit can be defined by combining base units algebraically. The SI is important because it is more than a set of distinct units. As the name implies, it is a system of units. The relationships among the various SI units, both basic and derived, are of great concern to those setting up and using a modern system of scientific measurements. The person who needs the SI units in a mathematical but primarily nonscientific application should still have some understanding of relationships.

Theoretically, many different units of mass, length, and time may be associated in definitions of derived units. A standard system often used is the meter-kilogram-second system, abbreviated **MKS**. In 1948 the International Conference defined the ampere in terms of the MKS system. The SI system is now based on the extended MKSA system associating the first four basic units. A second system still commonly used is the cent-gram-second system (with symbol **CGS**) proposed in Great Britain.

Example 1

Convert the expression 34 $g/cm^2/s$ from the CGS to the MKS system. In the expression, the slanted bar is read "per." 34 grams per cm^2 is 0.034 kg per cm^2, or 340 kg per m^2, so the new expression is 340 $kg/m^2/s$.

One great advantage of the MKS system for scientists is that it is *coherent* for both mechanical and electrical units. What this statement means is that derived units can all be obtained from the basic units without the introduction of numerical factors. Not even a power of ten is needed. The desire to have a coherent system is instrumental in deciding among possible definitions for various SI-derived units. For example, force could be measured in poundals or dynes, but only the newton is coherent in the SI. Although the average person using SI units may not need to understand the importance of a coherent system, he should realize that scientists who have developed the system consider the fact that it is coherent a great advantage of the modernized metric system.

The examples of derived units given below show how exponential notation can be used to provide very concise definitions, because of the fact that the system is coherent.

Example 2

The newton, although used in the definition of a basic unit (the ampere), is itself a derived unit defined by $1 \cdot N = 1 \cdot kg \cdot m \cdot s^{-2}$. This concise notation should be compared with the written definition appearing in Chapter 4. Note that s^{-2} is interpreted as per second per second, as it is used in connection with acceleration. Note also that the newton is defined in the MKS system (kilograms meter per second per second).

Example 3

The SI unit for pressure, the pascal (PA), can be defined by the equation $1 \cdot \text{Pa} = 1 \cdot \text{N} \cdot \text{m}^{-2}$. This definition may be expressed in words by saying that 1 Pascal is a unit of pressure measuring the force in newtons per square meter. This equation may be rewritten entirely in basic units by applying the definition of newton in Example 2.

$$\begin{aligned} 1 \cdot \text{Pa} &= 1 \cdot \text{kg} \cdot \text{m} \cdot \text{s}^{-2} \cdot \text{m}^{-2} \\ &= 1 \cdot \text{kg} \cdot \text{m}^{-1} \cdot \text{s}^{-2}. \end{aligned}$$

Many other derived units are appropriate for various branches of science, all defined in terms of basic SI units as part of a coherent system. But it should be kept in mind that the SI is not a complete, static system but one that is being changed and developed to meet new needs. For instance, a basic SI unit associated with the measurement of sound does not yet exist, so this may be a new basic unit in the near future. The following set of exercises should be considered as optional for the person who does not need to work with various derived units in science.

Exercise Set 15

For Exercises 1 to 4, rewrite each expression in the MKS system.

1. 5 g/cm/s
2. 6 $\text{g/cm}^2\text{/s}$
3. 8 $\text{g/cm}^2\text{/s}$
4. 14 $\text{g/cm}^3\text{/s}$

For Exercises 5 to 10, write in words the meaning of each definition.

5. $1 \cdot \text{J} = 1 \cdot \text{N} \cdot \text{m}$ (a joule, J, is a unit of work or energy)
6. $1 \cdot \text{W} = 1 \cdot \text{J} \cdot \text{s}^{-1}$ (a watt, W, is a unit of power)
7. $1 \cdot \text{V} = 1 \cdot \text{W} \cdot \text{A}^{-1}$ (a volt, V, is a unit of voltage)
8. $1 \cdot \Omega = 1 \cdot \text{V} \cdot \text{A}^{-1}$ (an ohm, Ω, is a unit of electrical resistance)
9. $1 \cdot \text{C} = 1 \cdot \text{A} \cdot \text{s}$ (a coulomb, C, is a unit of electric charge)
10. $1 \cdot \text{H} = 1 \cdot \text{V} \cdot \text{s} \cdot \text{A}^{-1}$ (a henry, H, is a unit of inductance)

For Exercises 11 to 14, rewrite each expression in terms of basic SI units only.

11. $1 \cdot \text{J} = 1 \cdot \text{N} \cdot \text{m}$
12. $1 \cdot \text{V} = 1 \cdot \text{W} \cdot \text{A}^{-1}$
13. $1 \cdot \text{W} = 1 \cdot \text{J} \cdot \text{s}^{-1}$
14. $1 \cdot \Omega = 1 \cdot \text{V} \cdot \text{A}^{-1}$

Mathematics Laboratory Activities 17

1. Study a list of derived SI units. Write the definition for several units as an equation in terms of the basic units only.
2. Study recent reports of changes suggested in SI units as a result of international conferences.
3. Find examples showing that the CGS system is not a coherent system of measurement.

BIBLIOGRAPHY

American National Standards Institute, *A Guide to the Use of SI – the International System of Units,* Z 210.1. American National Standards Institute, 1971.

Chiswell, Barry, and E. C. M. Grigg, *SI Units.* New York: John Wiley and Sons Australasia, 1971.

Committee on the Metric System, *The Metric System of Weights and Measures,* 20th Yearbook of the National Council of Teachers of Mathematics. New York: Bureau of Publications, Columbia University, 1948.

Danloux-Dumesmils, Maurice, *The Metric System, a Critical Study of its Principles and Practice,* translated from the French by Anne Garrett and J. S. Rowlinson. London: Athlone Publishers, 1969.

Davies, Charles, *The Metric System.* New York and Chicago: A. S. Barnes and Company, 1871.

De Simone, Daniel V., *A Metric America: A Decision Whose Time Has Come.* National Bureau of Standards Special Publication 345. Washington, D.C.: U.S. Government Printing Office, July 1971.

Diamant, R. M. E., and P. Godfrey, *Understanding SI Metrication.* London, England: Angus and Robertson, 1970.

Donovan, Frank Robert, *Prepare Now for a Metric Future.* New York: Weybright and Talley, 1970.

Green, Marvin H., *International and Metrical Units of Measurement.* New York: Chemical Publishing Co., Inc., 1961.

Huntoon, Robert D., et al., *International Standards, an Interim Report of the U.S. Metric Study,* National Bureau of Standards Special Publication 345-1. Washington, D.C.: U.S. Government Printing Office, December 1970.

Jones, M. J. B., *A Guide to Metrication.* Oxford, England: Pergamon Press, Ltd., 1969.

Le Maraic, A., *Two-Way Conversion Tables of Units of Weights and Measures, Metric and United States Customary.* Greenwich, Connecticut: Abbey Books, 1953.

National Bureau of Standards, *International System of Units (SI),* NBS Special Publication 330. Washington, D.C.: U.S. Government Printing Office, 1972.

National Council of Teachers of Mathematics, *The Arithmetic Teacher,* Vol. 20, No. 4, April 1973, and No. 5, May 1973.

National Industrial Conference Board, *The Metric Versus the English System of Weights and Measures.* New York: Century, 1921.

Palmer, James F., *The International System of Units (SI) Handbook.* Huntsville, Alabama: Brown Engineering Co., 1969.

Ramsay, D. C., and G. W. Taylor, *SI Metrication: Easy to Use Conversion Tables, British and U.S to Metric (SI) Units.* Edinburgh: W. and R. Chambers, 1969.

Ronningen Metric Company, *Weights and Measures, U.S. to Metric, Metric to U.S.* Phoenix: Sims Printing Company, 1969.

United States Bureau of Standards, *The International Metric System of Weights and Measures.* Washington, D.C.: U.S. Government Printing Office, 1932.

United States Department of Commerce, *Brief History of Measurement Systems.* Washington, D.C.: U.S. Government Printing Office, Special Publication 304A, October 1972.

United States Department of Commerce, *A Metric America,* Reprinted from National Bureau of Standards, Technical News Bulletin, September 1971.

United States Department of Commerce, *Units of Measurement: Conversion Factors and Special Tables,* October 1972. Reprinted from *Units of Weight and Measure – International (Metric) and U.S. Customary,* NBS Miscellaneous Publication 286, May 1967.

ANSWERS TO SELECTED EXERCISES

Exercise Set 1. *1.* Conversion within the metric system is based on a decimal notation of powers of ten, but conversion within the English system depends on various fractions. The basic units of the metric system involve standards from nature rather than standards derived from measurements of the human body. *3.* France *5.* A gram was the mass of one cubic centimeter of water at its temperature of maximum density. *7.* piles of cordwood *9.* International System of Units *11.* Mouton's proposed standard of length

Exercise Set 2. *1.* 128 *3.* 100,000 *5.* 1/81 *7.* 1/1,000 *9.* 100 *11.* 1,000,000 *13.* 0.1 *15.* 0.001 *17.* 0.000,000,001

Exercise Set 3. *1.* $1 \text{ m} = 10^{-2}$ hm *3.* $1 \text{ m} = 10$ dm *5.* $1 \text{ m} = 10^3$ mm *7.* 430 m *9.* 0.243 m *11.* 4380 m *13.* 0.032 km *15.* 320 dm *17.* 32,000 mm *19.* 4.52×10^{-4} km *21.* 452 mm *24.* 4.52×10^{-2} dam *25.* 3.457×10^{-5} km *27.* 3.457×10^{-3} dam *29.* 2.38×10^4 cm *31.* 10^{-10} m *33.* 9.46×10^{15} m

Exercise Set 4. *1.* 4.3 cm *3.* 74.4 m *5.* 47 *7.* 746 m/min *9.* 185 km *11.* 0.374 cm *13.* 1.5 km *15.* 14.468 cm

Exercise Set 5. *1.* $\approx$ 11 m *2.* $\approx$ 3 yd *5.* $\approx$ 411 km

7. ≈ 38 cm *9.* ≈ 104 km/hr *11.* ≈ 40 km/hr
13. ≈ 16 km/hr *15.* ≈ 60 cm *17.* ≈ 90 cm
19. 17.78 cm *21.* 3.6576 m *23.* 15.748 cm

Exercise Set 6. *1.* 2.6×10^7 m^2 *3.* 4.2×10^3 m^2
5. 8.63×10^{-2} m^2 *7.* 2.1×10^{-1} dam^2
9. 3.74×10^{-2} hm^2 *11.* 2.5×10^4 mm^2
13. 3.5×10^3 hm^2 *15.* 3.5×10^7 μm^2
17. 2×10^7 cm^2 *19.* 14.1 ha *21.* ≈ 0.56 cm^2
23. ≈ 23 cm *25.* ≈ 814 km^2 *27.* ≈ 9.1 km^2

Exercise Set 7. *1.* ≈ 3 sq in. *3.* ≈ 42 sq yd *5.* ≈ 86 sq mi
7. ≈ 575 acres *9.* ≈ 107 cm^2 *11.* ≈ 1 sq in.
13. 45.1612 cm^2 *15.* 12.54191040 m^2

Exercise Set 8. *1.* 1.9×10^9 m^3 *3.* 2.5×10^{-5} m^3
5. 1.24×10^{20} m^3 *7.* 7.54×10^7 m^3 *9.* 5×10^{-3} dam^3
11. 1.28×10^{-7} km^3 *13.* 4.2×10^5 cm^3
15. 3.45×10^5 hm^3 *17.* 3.27×10^{-3} cm^3 *19.* $7.74
21. 79.0% *23.* 149 cm^3 *25.* ≈ 357 m^3

Exercise Set 9. *1.* ≈ 1225 cu ft *3.* ≈ 394 cm^3
5. ≈ 4 liters *7.* ≈ 45 qt *9.* ≈ 1360 cm^3 *11.* ≈ 47¢
13. ≈ 550 cm^3 *15.* ≈ 180 ml *17.* ≈ 10 ml
19. ≈ 1.25 ml

Exercise Set 10. *1.* 3.4×10^4 g *3.* 6×10^3 mg
5. .350 g *7.* .9427 kg *9.* 7.42×10^3 μg *11.* $\frac{1}{4}$
13. $\frac{1}{5}$ *15.* 2.28 t *17.* 10^9 kg *19.* 1.5×10^4 g/km^3
21. 3.1% *23.* The 650-g box at 79¢

Exercise Set 11. *1.* ≈ 336 g *3.* ≈ 68 kg *5.* ≈ 11 ton
7. ≈ 104 g *9.* ≈ 0 lb *11.* ≈ 5 ton *13.* ≈ 588 g
15. ≈ 224 g *17.* ≈ 64¢ *19.* ≈ 17¢ *21.* ≈ 3.4 kg
23. 1.1 kg

Exercise Set 12. *1.* 76.85°C *3.* − 209.65°C *5.* ≈ 64°F
7. ≈ 19°F *9.* ≈ 80°F *11.* ≈ 21°C *13.* ≈ − 7°C
15. ≈ 177°C *17.* 86°F *19.* 50°F *21.* ≈ 190°C
23. ≈ 38.6°C *25.* 96

Exercise Set 13. *1.* 35,000 m/s *3.* 0.15 m/s *5.* ≈ 4.5 m/s *7.* ≈ 3.9 m/s *9.* 299,792,500 m/s *11.* 299.7925 km *13.* 3,400 A *15.* 0.382 kA *17.* 20π

Exercise Set 14. *1.* $\pi/4$ *3.* $3\pi/8$ *5.* ≈ 0.9993 *7.* ≈ 3.1291 *9.* ≈ 1.7183 *11.* $\pi/6$ *13.* $2\pi/3$ *15.* $3\pi/2$ *17.* 225° *19.* $112\frac{1}{2}°$ *21.* 2π *23.* 0.4π

Exercise Set 15. *1.* 0.5 k/m/s *3.* 80 kg/m²/s *5.* One joule is the work done by a force of one newton through a distance of one meter. *7.* One volt is a unit of voltage equal to one watt per ampere. *9.* One coulomb is an electrical charge of one ampere for one second. *11.* $1 \cdot \text{J} = 1 \cdot \text{kg} \cdot \text{m}^2 \cdot \text{s}^{-2}$ *13.* $1 \cdot \text{W} = 1 \cdot \text{kg} \cdot \text{m}^2 \cdot \text{s}^{-3}$

INDEX